城市行走书系
策划:江岱、姜庆共

布拉格建筑地图
中文:蒲仪军
英文:埃兰·诺依曼·菲仕乐
摄影:蒲仪军、埃兰·诺依曼·菲仕乐、李辉、刘奚、贝内迪克特·马克尔、菲利普·什拉帕尔,等

责任编辑:金言
书籍设计:张微

CityWalk Series
Curator: Jiang Dai, Jiang Qinggong

A Passage Through Prague Architecture
Chinese Text: Pu Yijun
English Text: Elan Neuman Fessler
Photograph: Pu Yijun, Elan Neuman Fessler, Li Hui, Liu Xi, Benedikt Markel, Filip Šlapal, et al.

Editor: Jin Yan
Book Designer: Zhang Wei

同济大学出版社
Tongji University Press

A Passage Through Prague Architecture
布拉格建筑地图

蒲仪军　［捷］埃兰·诺依曼·菲仕乐　著
Pu Yijun　Elan Neuman Fessler

同济大学出版社·上海

序言

城市作为人类聚居的空间领域，是各种文明存在和演化的物质见证和精神载体，留下了异彩纷呈的建成遗产，包括城池、道路、广场、街区、建筑、景观等构成要素，其中可反映出生活方式、行为习惯、价值取向、美学气质等人文气息与场所精神，及其变化轨迹。捷克共和国的首都布拉格，就是这样一座历史悠久、遗产璀璨的中欧名城，被誉为欧洲建筑博物馆。

布拉格位于欧洲地理位置的中心，因未在第二次世界大战中受到大规模破坏，基本上完整地保留了中世纪以来各时期多种风格的建成遗产及城市风貌，因而成为全球城市中第一座被联合国教科文组织整体列入"世界遗产"名录的城市。如今，布拉格的建成遗产已经成为其在可持续发展中城市历史身份和价值无可替代的文化资源。而布拉格作为世界级文化旅游目的地首选之一的主要缘由亦在于此。

《布拉格建筑地图》一书，是同济大学出版社城市行走书系中第一本涉及国外名城的著述。书中通过43幢不同时期历史建筑的展示，解读了布拉格的城市特色及变迁特征。尽管价值连城，这些建成遗产却大多在西方建筑史研究领域关注的重点之外，因而其地位和意义有必要关注和重估，其保护、传承和活化的方法和经验，也值得国内城市遗产界重视和参考。

本书为中捷两位作者合著，其中上海的蒲仪军博士曾经在布拉格访学，近年来通过实地踏勘、资料检索和论文撰写，做出了拓展性成果。相信本书的出版，不仅可以作为布拉格的建筑导览，而且也有益于我国城市建成遗产保护事业。

是为序。

同济大学建筑与城市规划学院教授
中国科学院院士
辛丑冬月书于沪上寓所

Foreword

As physical spaces inhabited by human beings, cities are the material witness and spiritual carrier of the existence and evolution of various civilizations, leaving colorful built heritages including city walls, roads, squares, blocks, architectures, landscapes and other constituent elements. They can reflect the humanistic atmosphere and genius loci such as lifestyles, habits, value orientation and aesthetics, as well as their trajectories of change. Prague, the capital of the Czech Republic, is a famous city in Central Europe with a long history and brilliant heritage as such. It is known as the European architectural museum.

Prague is located at the center of Europe. Having escaped from severe damages in World War II, Prague basically retains the built heritage and urban landscapes of various styles since the Middle Ages. Therefore, it has become the first city in the world to be listed as a "World Heritage" by UNESCO as a whole. Today, Prague's built heritage has become an irreplaceable cultural resource representing the urban historical identity and values in the sustainable development of this city. This is also the main reason why Prague is one of the first choices among world-class cultural tourism destinations.

A Passage Through Prague Architecture is the first book in the *CityWalk* Series of Tongji University Press that involves famous foreign cities. Through the display of 43 historic buildings from different periods, this book interprets Prague's urban characteristics and changes. Though priceless, most of these built heritages are beyond the focus of western architectural history research, and their status and significance need to be readdressed and reassessed; The methods and experiences in their conservation, inheritance and revitalization are also worthy of attention and reference by the domestic urban heritage community.

This book is co-authored by a Chinese and a Czech scholar. One of them, Dr Pu Yijun, who is from Shanghai, has the experience of academic exchange in Prague and has made extensive academic achievements through fieldwork, data retrieval and thesis writing in recent years. I believe that the publication of this book can not only serve as an architectural guide of Prague city, but also benefit the conservation of urban built heritage in China.

Chang Qing

Professor at the College of Architecture and Urban Planning, Tongji University

Academician of Chinese Academy of Sciences

目录

序言 ································ 6
请先阅读 ···························· 12
布拉格简介 ························· 16
背景：从海底火山到斯拉夫居住地 ··· 22

罗马风时期 ······················· 28
 高堡的圣马丁圆形小教堂 ········· 34
哥特时期 ························· 40
 犹太区老 - 新犹太教堂 ············ 48
 布拉格城堡内的圣维特大教堂 ····· 56
 查理大桥 ························· 64
 老城广场的旧市政厅 ·············· 68
文艺复兴时期 ···················· 76
 安娜王后的夏宫 ·················· 84
 施瓦岑贝格宫 ···················· 88
巴洛克时期 ······················ 92
 华伦斯坦宫和花园 ··············· 100
 克拉姆 - 葛拉斯宫 ················ 104
 圣尼古拉斯教堂(小城) ·········· 108
 斯特拉霍夫修道院及其图书馆 ···· 112
古典复兴时期 ··················· 116
 城邦剧院 ························ 124
 金斯基花园 ······················ 128
 马萨里克火车站 ················· 132
 国家大剧院 ······················ 136
 鲁道夫音乐厅 ···················· 140
 布拉格中央火车站 ··············· 144
 市民会馆 ························ 148
现代主义时期 ··················· 152
 卢塞纳宫 ························ 160
 黑色圣母屋 ······················ 166

亚德里亚宫 ······················· 170
展览会宫 ························· 174
拔佳鞋屋 ························· 178
马内斯美术馆 ···················· 182
芭芭住宅区 ······················· 186
缪勒别墅 ························· 190
耶稣圣心教堂 ···················· 194
共产主义时期 ··················· 198
 国际饭店 ························ 206
 斯拉夫本笃会修道院(修复) ······ 210
 联邦议会大厦 ···················· 214
 Máj 百货公司 ···················· 218
 布拉格会议中心 ················· 222
 新舞台 ·························· 226
 日什科夫电视塔 ················· 232
当代 ···························· 236
 会跳舞的房子 ···················· 242
 金色天使大厦 ···················· 246
 皇家花园的柑橘暖房 ············· 250
 鹿谷步道 ························ 254
 欧洲大厦 ························ 258
 国家技术图书馆 ················· 262
 DRN/ 民族大厦 ·················· 266
 DOX 当代艺术中心 ·············· 270
 布拉格艺术馆 ···················· 276
新的未来 ······················· 282

建筑名录 ·························· 286
参考文献 ·························· 290
后记一 ···························· 292
后记二 ···························· 294

Contents

Foreword ·· 7
Preamble ·· 14
Introduction to the City ····················· 20
Background: From a Volcanic Sea-World to the Slavic Settlements ········ 26

THE ROMANESQUE ···························· 32
St. Martin's Rotunda at Vyšehrad Castle ··· 34
THE GOTHIC ·· 44
Old-New Synagogue of the Jewish Town ··· 48
St. Vitus Cathedral in Prague Castle ····· 56
Charles Bridge ······································ 64
Old Town Hall ·· 68
THE RENAISSANCE ·························· 79
Queen Anne's Summer Palace ············ 84
Schwarzenberg Palace ······················· 88
THE BAROQUE ······································ 96
Wallenstein Palace and Gardens ······ 100
Clam-Gallas Palace ···························· 104
St. Nicholas Church (Lesser Town) ······ 108
Strahov Monastery and Its Library ······ 112
THE REVIVALIST TIME ···················· 122
The Estates Theatre ·························· 124
Kinský Folly and Garden ···················· 128
Masaryk Railway Station ···················· 132
National Theatre ································ 136
Rudolfinum ·· 140
Prague Main Train Station ················ 144
Municipal House ································ 148
THE MODERNIST TIME ···················· 158
Palace Lucerna ·································· 160
House of the Black Madonna ············ 166

Adria Palace ·· 170
Trade Fair Palace ······························· 174
Baťa House (Palace) of Shoes ········· 178
Mánes Union of Fine Arts ·················· 182
Baba Residential Estate ···················· 186
Villa Müller ··· 190
Church of the Most Sacred Heart of Our Lord 194
THE COMMUNIST TIME ···················· 202
Hotel International ····························· 206
Emauzy Convent ································ 210
Federal Assembly Building ················ 214
Máj Department Store ······················ 218
Prague Congress Centre ··················· 222
New Scene of National Theatre ········ 226
Žižkov Television Tower ····················· 232
CONTEMPORARY ································ 239
Dancing House ··································· 242
Golden Angel ······································ 246
Orangery in the Royal Gardens ········· 250
Deer Moat Passage ··························· 254
Palace Euro ··· 258
National Technical Library ················ 262
DRN / Národní Palace ························ 266
DOX Centre for Contemporary Art ······ 270
Kunsthalle Praha ································ 276
THE NEAR FUTURE ··························· 284

Individual Building Table ··················· 288
Reference ·· 290
Afterword 1 ·· 293
Afterword 2 ·· 295

请先阅读

人类社会和思想随着历史浪潮不断前行,在不同的历史时期产生了不同风格的建筑。今天各个时期的建筑都在当下与布拉格这座城市共存,并继续强化着这座波希米亚城市的独特性。本书通过划分建筑发展的历史时期、分析典型建筑,展示出布拉格一千年来在地质、地貌、历史事件以及人的生活观念等多重因素持续相互作用下所产生的建成环境。

书中讨论了布拉格这座城市中多种风格的建筑,着重关注它们在时间和空间上对这样一座有机生长的城市所产生的意义。采用合成分析法、分层分析法或全息分析法对这一复杂主题进行研究;选择建筑时考虑它在历史长河中所处

建筑地图(图中序号表示本书所选 43 幢建筑,具体可见建筑名录)
The Map of Buildings (The serial numbers in the figure indicates the 43 buildings selected in this book. See the list of buildings for details.)

的位置、建成环境的特征，以及它对城市意识形态的影响。通过横向与纵向的动态透视，对这些建筑进行全方位审视，旨在邀请读者从一种更广阔且多方位的视角来观察布拉格城市内部的建筑景致。

本书试图采用整体与积累观察的手段，通过个体分析与分层分析的方法，解读布拉格的内部建筑力量与结构。没有任何一幢建筑是独立存在的，建筑的周边环境应当被考虑为建筑的一部分，因为它以一种历史的、物质的和精神性的形式，不可避免地包含了过去和现在，也超越了自身的存在。

作者希望本书可以使读者对布拉格这座城市、城市中的建筑及其隐藏结构的复杂性有所认知，并由此对欧洲的历史变迁有所理解，进而使得波希米亚特征可以被读者所理解，并且向大家阐释"神奇布拉格"是什么，以及如何实现。

书中构建了阅读每一幢选定建筑的方式，以及基础要素：土地、人民、市场、法律和神权。每一种要素同样也可以被看作是构建城市景观系统必要的社会及物质元素。城市内部交错编织的矩阵式结构正是由这些不断积累的历史和社会结构叠加产生的。随着这些不同的社会元素开始在这个矩阵中书写、建立自己的特定历史，它们之间的历史分层和空间密度变得越来越复杂。

本书筛选了43幢布拉格建筑。每一幢建筑都将其自身实体以及象征属性根植在城市景观中。这些建筑为读者更好地去阅读理解这座城市，并同时在这个复杂的城市结构中去了解波希米亚特征提供了可能。书中对建筑发展的历史时期进行了划分，确定为罗马风、哥特式、文艺复兴、巴洛克、古典复兴、现代主义、共产主义和当代等不同时期建筑风格，并针对每种风格选取了具有代表性的建筑。在考虑这些建筑在城市中历史地位的同时，将它们放在相应章节中，使得在详细描述每栋建筑的特定"故事"之前，可以概述其所处的历史时期和所经历的风云变幻，并按照时间顺序追溯布拉格一千年来的社会历史变迁。

本书融汇了构成布拉格整体结构的意识形态、空间、符号以及社会等层级要素。在这个城市空间中，历史的各个层级都在持续向前迈进。这些建筑不仅是历史的提醒者，也是这个城市舞台上城市景观剧的表演者。需要注意的是，考虑到东西方不同的阅读习惯和思维模式，书中的中英文其实是相互参照与补充的，希望从不同的视角为读者提供更多信息。

Preamble

The purpose of this book is to inform about architecture. It is also about the history of society and the history of thought that has produced architecture at different times. Since various architectural styles at all times co-exist within the contemporary city of Prague, their collective impact continues to influence the particularities of this Bohemian city. A Passage Through Prague Architecture reveals the continuing material consequence of the interaction of forces such as geography, topography, historical events, and lives and ideas of people over time.

We will discuss the many architectural styles found in the city, with attention to the meanings they have brought (in time and space) to this organic tapestry. The approach to this complex subject might be called synthetic, layered or holographic. By the examination of each building, three possibilities are considered at once: its position within a historical continuum, its three dimensional situation within its site, and its ideological impact. The review moves simultaneously through the city both horizontally and vertically. This multi-layered perspective, which spans a broad sweep of time, is meant to invite our visitors to look at Prague's sights and sites from a wider angle and with a multifaceted point of view.

This approach is an attempt to capture forces and structures simultaneously, in their individuality and through their layers, in a holistic and cumulative manner. It is easy to forget perhaps that no structure stands in isolation. Even the "empty" space surrounding a building must be considered as a part of the built structure, since it inevitably has both a past and a present — a historical, material and symbolic form — and a presence which engages its place beyond itself.

We hope this book will help to explain the complexities of the city of Prague, its architecture and its hidden structures, and make intelligible the movements of European histories through which these architectural productions and the reaction to them have emerged and can be interpreted. More specifically, we hope to make the Bohemian character understandable for our readers, and to suggest for all, the what and how of what is commonly perceived as the magic of Prague.

We have identified fundamental elements or categories that frame the way of reading each specific building: the land, the people, the markets, the law and the sacred. Each element can also be seen as a social and material piece of the fabric that necessarily makes up the landscape of the city. The cross-referenced and interwoven matrix of the city is produced by the overlay of these elements and their influences accumulated over time.

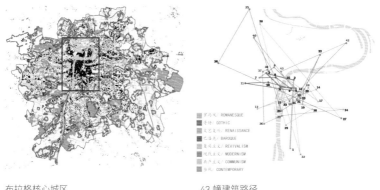

布拉格核心城区
The Core Area of Prague City

43 幢建筑路径
A Passage Through 43 Buildings

The density of both the historical layers and the spaces of the city between them becomes increasingly complex as these differing elements begin to establish their own specific histories within this matrix.

We have selected forty-three buildings/institutions in Prague. Each inserts both physically and symbolically, its "idea" into the city landscape. Together these examples compose a possible reading of the city, and of the configuration of its "Bohemian character" — in its complexity. We have subdivided historical time in this book and selected the 43 structures in line with emblematic moments in architecture already identified as: Romanesque, Gothic, Renaissance, Baroque, Revivalist, Modernist, Communist, and Contemporary. The placing of these buildings — inseparable from their historical positions in the city — into chapters that correspond to the above mentioned "styles" allows us to present an overview of the moments, cycles and surges of history before detailing the particular "story" of each building, and to track in chronological order the historical and social transformations of the city over a period of 1000 years.

This book interlaces the layered web of ideology, space, symbols and society in the structures that make up Prague. In the space of the city all layers of history continue. They are not merely reminders, but living actors in the landscapes of the drama still playing itself out upon the city's stage. The Chinese and English text of this book complement each other with cross references, considering the differences of reading habits and patterns of thought between East and West, in the hopes of giving readers more information from different perspectives.

布拉格简介

捷克共和国首都布拉格位于欧洲地理位置的中心，拥有千年的历史。这座城市几乎完整保存着从中世纪到当代各个时期和多种风格的建筑及风貌，战争并未对其造成实质性的破坏，因此被誉为欧洲建筑文化博物馆。彼得·沃尔利克认为布拉格不同时期的历史痕迹在其城市空间中共生的现象令人叹为观止，并援引克里斯蒂安·诺伯格-舒尔茨著名的《场所精神》，称赞它是"少数几个'健康'自然、层次丰富，同时具有内在精神的环境典范之一"。布拉格的历史城区在 1992 年被列入世界文化遗产名录。

布拉格不仅仅是建筑艺术博物馆，也是稀有的地理博物馆。它的主体坐落于群山包围的盆地中，大范围的城市区域内有七座山丘，其中维舍赫拉德山（高堡）、赫拉德坎尼山成为形成布拉格中心盆地地形的主要因素，城市的主体夹在两山之间。伏尔塔瓦河是布拉格最显著的地景元素，是控制城市发展的天然轴线。它作为布拉格城市空间的天然的主脊，重塑着地景，在城市结构的发展中扮演重要的角色。1874 年，贝德里赫·斯美塔那为赞颂这美丽的自然景观，

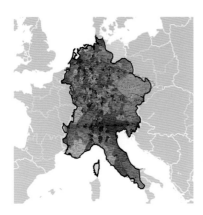

神圣罗马帝国，1070
Holy Roman Empire, 1070

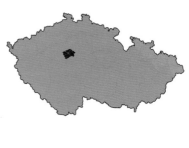

捷克共和国，1993
Czech Republic, 1993

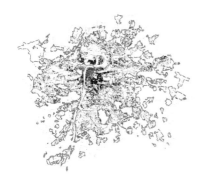

布拉格城区，2019
City Of Prague, 2019

布拉格历史中心，2019
Historical City Center, 2019

创作了著名的交响乐史诗《我的祖国》。这种起伏变化的城市地貌使布拉格空间层次丰富，也成为布拉格城市魅力的重要组成部分。

　　布拉格历史悠久，多样而包容。从旧石器时代起，这一地区就有人类定居。6世纪，一支斯拉夫部落由西而入，定居在波希米亚地区，他们就是捷克民族的祖先。布拉格是以9世纪博日沃伊一世在伏尔塔瓦河右岸建造的高堡为核心，然后逐渐成形，后来他在河对岸建造了另一座城堡，也就是今天的布拉格城堡。不久后，布拉格成为波希米亚的首都，并成为欧洲南北商路上一个重要的贸易中心，波希米亚也于1002年加入神圣罗马帝国。像欧洲许多前工业化的中世纪城市一样，布拉格在城市发展过程中经历了四个主要时期：中世纪教会与市民力量的抗争、外国势力的介入、战争的冲击和工业化时期的城市革命。也由此逐渐形成四个主要的区域：伏尔塔瓦河以西的城堡区、小城、伏尔塔瓦河以东的旧城区和新城区。19世纪以来,由于特殊的地理位置,布拉格和波希米亚(捷克的别称)成为欧洲历史上一个重要的舞台，先后经历了奥匈帝国、第一共和国、纳粹占领、共产主义和后共产主义时期。历史上的每一个时期都在布拉格城市空间中留下了印记，各种风格的建筑在城市中和谐共生。作为欧洲的地理中心，斯拉夫、日耳曼、法兰西和意大利等周边文化的影响也共同汇聚到布拉格，在包容和吸收的同时，布拉格成为世界性和地方性统一的城市。

布拉格老城是一本千年的建筑风格演变的教科书，汇集了从中世纪开始的各种风格：罗马风、哥特式、文艺复兴、巴洛克、洛可可、古典复兴、新艺术运动、装饰艺术、立体主义、功能主义等。这些不同时期及风格的建筑宁静祥和地矗立在一起，丝毫未受到现代世界打扰，时间也仿佛在这里驻足。尽管布拉格历史城区的大多数建筑主要建造于19世纪和20世纪之间，但城市的肌理依旧保持着最初的样子。将1940年和2010年老城区的照片进行叠合对比，80%的地方可以重合。不同的时代痕迹有深有浅，但是没有一个时代可以成功地占据整个风貌。历史上，布拉格经历了多次战争，特别是两次世界大战，但其城市风貌却没有遭到巨大的破坏，保持了完整性和原真性，这在欧洲城市中是罕见的。这与捷克的民族性密切相关，对一个要独自承受自身命运的人和承受自身历史的民族来说，忍耐和不屈不挠是必须的。一座城市，也必须有这样的素质。在捷克语中，像在许多其他语言中一样，"忍耐"这个词和动词"遭受"有着同样的词根。从表面上看这座城市免遭了战争的蹂躏，然而比起那些直接受交战影响的城市这座城市却忍受得更多。

布拉格环境优美，城市依地形而建，地貌及空间尺度宜人，视觉景观变换丰富，这种视觉景观的丰富性随季节和气候变化而成倍增长。行进中良好的空间感受也成为外国旅行者甚至专业人士对布拉格最深印象之所在。伏尔塔瓦河的水平与环抱山丘的垂直对比；历史城区连片的红瓦屋面和无数突出的教堂钟塔对比；哥特式的圣维特教堂和超现实主义的日什科夫电视塔对比：共同构筑了布拉格独特的场所特征。尼采说："当我想以一个词来表达音乐时，我找到了维也纳；而当我想以一个词来表达神秘时，我只想到了布拉格。"这种神秘意象也在布拉格人卡夫卡的小说中呈现。哥特式的城市空间及巷道成为多部电影的取景地。街道、塔楼、教堂、大桥；空间开合、转折、迷失；清晨、黄昏、深夜：引导人进入一个个无止境的内部。古老街区的空间总让人有未知的感觉，既神秘又可怕，既温暖又安全。

从最早的人类定居点到罗马风和哥特式的教堂，从文艺复兴和巴洛克式时期的城市宫殿到第一共和国时期的郊野功能主义别墅，从共产主义时期的城市副中心到当代城郊居住区，布拉格的故事就此展开。

Introduction to the City

At first glance the city appears to be a unified, if complicated, whole. Its image, structure and form, however, have come into being over time, as much by accident and circumstance as by coordinated plans and design. This rich urban centre, over 10 centuries old, was designated as a UNESCO city of world heritage in 1992.

From the beginning, different settlements and claims on the land necessitated law. For as long as the land was settled, there existed a rule over it. In 870 the Přemyslid family founded the Prague Castle. Royal families maintained control and order over the people and the various activities pertaining to the land. When Bohemia joined the Holy Roman Empire in 1002, an added sacred order was institutionalised under the patronage of nobilities. The markets, places of political power, action and social exchanges, would progressively yield new productions through crafts, enterprise and capital. The many different and specialised people in these markets lived within the security of the two higher laws of the land: the holy and the royal.

Prague is the central region where three rivers, Vltava, Berounka and Sazava meet. In 1874, Bedřich Smetana composed a now-famous musical piece, *Má Vlast (My Homeland)*, in celebration of this natural space. From the edges of the region, through the greater landscape of the city into the expressive curves of the river basin on the variegated slopes beneath the Castle — here is the locus of Bohemia. The larger region, bordered by mountains, is a natural fortification defining the limits of Bohemia for over a thousand years. Within the centralised — yet not quite smooth topography of the region — the city itself is echoed.

Prague emerged within this natural form of ancient geology — valleys, forests and hills, extending upwards into mountains. Passing through the centre of Prague, the Vltava flows northwards down the river valley to the northern German sea. To varying degrees, its waters are sourced from the Austrian hills in the south, the Moravian streams in the east, and the forested hills in the west. Prague, as the centre of Bohemia, is also the centre of Europe. The long history of Germanic, Slavic and Italian influence on the architecture of Prague is pre-figured to some degree by this geographical situation at the heart of the continent.

From Prince Bořivoj and the Přemyslid dynasty — when the first institutions of today's society and urban territory were established — to the hills, valleys and streams of early neolithic tribes, through an ancient geological process of the landscape's topological formation, we can retrace the story of today's Prague. From the contemporary urban agglomerations on the city's outskirts, to suburban centres of the Communist period, past modernist villas on the outer hills of the First Republic and an expanding 19th century city core, to the various Baroque and Renaissance urban estates and palaces, through Gothic and Romanesque enclaves to its earliest beginnings, the story of today's Prague unwinds.

Woven through the Romanesque, Gothic, Renaissance, Baroque, Revivalist, Modernist, Communist, and Contemporary times, Prague today, has 21st century interiors built upon 12th century cellars. While most buildings in Prague date primarily from the 19th and 20th centuries, many of the city's streets remain within their ancient contoured pattern, and these early paths and waterways still affect the structure, flow and life of the land.

背景：从海底火山到斯拉夫居住地

城市与传说

欧洲大多城市的诞生故事都源于神话传说。布拉格也有一个象征之地、一幢建筑的故事，它后来所有与建城相关的思想都建立在这个故事之上。在罗马式建筑首次在这片土地上留下印记之前，它的轮廓已经建立起来……

城市

维舍赫拉德山（高堡）、赫拉德坎尼山和维特克夫山，这三座山连接成一个神圣三角，就是波希米亚的中心。高堡是皇室的第一个定居地，位于布拉格以南，是伏尔塔瓦河的守护者；赫拉德坎尼山，是城堡的所在地，塔楼高耸，是太阳的守护者；位于城外的维特克夫山，是大地的守护者：它们共同孕育了这片土地。

布拉格

布拉格，捷克语的意思是门槛，它的含义包括空间和时间两个方面。布拉格在历史长河中发生了剧烈的变化，但它以重构的方式保留了其古老且永恒的特征。在每个历史时期，它都可以适应并吸收变化的力量。布拉格本身就是一幢历史建筑，随着时间的推移，不断重新融入这片土地并被社会重新定义，它永远处于复兴的门槛上。

土地（数百万年前）

土地是布拉格形成的重要因素。布拉格曾经是水下火山的一部分，逐渐消退的水域刻画出土地的年轮，形成了山谷和丘陵。这些古老的地质背景仍继续影响着城市的布局。

人民（五千年前）

人是构成这片定居地的重要因素。从新石器时代山丘上集聚、史前道路的

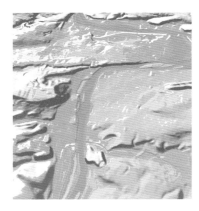

布拉格
Prague (Praha)

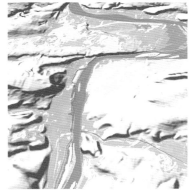

土地
The Land

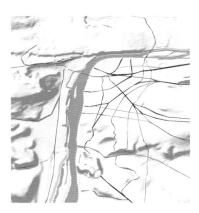

人民
The People

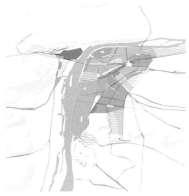

市场
The Markets

开辟到斯拉夫部落的建立,这些临时聚落逐渐对地形和地貌产生影响,也促进了木材和石材的利用。人们通过伐木、狩猎、采集等行径的自然路线将各种贸易中心联系起来。

法律
The Law

神权
The Sacred

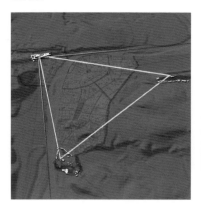

城市
The City

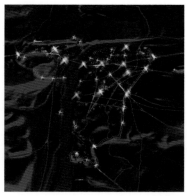

混合
The Hybrid

市场(800年)

　　市场作为一种社会经济因素,组织着土地上人的活动。地形首先决定了市场和定居点的位置,从而决定了社会等级结构和权力在景观中的位置。布拉格最大、最繁华的贸易地点位于伏尔塔瓦河湾旁的平坦地,这一区域的中心日后发展成为老城广场。

法律（870 年）

法律确立了土地上的秩序并确保了社会安全。集市、城门、塔楼、教堂和其他由石头和木材建造的房子都被城墙包围并保护着。城墙代表了一种权力机构，围合出社区。君主在城堡里统治这片土地，通过法律对领土、市场、市民的集体安全进行保护。

神权（973 年）

神权赋予土地更高的意义。城市的神权在修造城市景观过程中扮演着一个变革的角色。自 973 年布拉格主教辖区成立以来，神权通过教堂、教会、修道院和城堡的建造不断强化着自身，但它始终服从于更大的欧洲秩序，一种象征性的法则组织着个人与世界的联系。

混合（1348 年）

神权、地貌和社会混合在一起，共同构成这座城市的空间结构。拓扑、历史和宇宙观融为一体，成为这座城市多层次的根基，并深深根植于此。布拉格的建筑让城市故事变得栩栩如生。

Background: From a Volcanic Sea-World to the Slavic Settlements

City and Myth

All cities are born to a myth. The myth tells a story of a symbolic ground, a construct, upon which all later thoughts of the city are based. Before the Romanesque made its first marks on the land, the skeleton of the city was already established…

The City

Vyšehrad, Hradčany and Vítkov are three founding hills. They establish the boundaries of a sacred triangle, the heart of Bohemia. This triangle is given both cosmological and topological form by these natural figures of the landscape. Vyšehrad, guardian of the river, on the south of Prague, became the first royal settlement. Hradčany, guardian of the sun, and seat of the Castle, towers over the land. Vítkov, the silent hill, outside the ancient city, is the geological and symbolic guardian of the earth.

Prague (Praha)

"Práh", in Czech, means threshold. In the case of this city, the threshold encompasses both space and time. Neither here nor there, it is forever caught in the space of transition. The city has changed throughout its history yet it has retained — in reconfigured forms — its ancient and enduring character. In itself, overlooked and indeterminate, the threshold, like the city, becomes a memory again with every passing moment. Re-inscribed into the land and redefined by society, Prague — itself as a historical building — remains perpetually on the threshold of its renewal.

The Land (millions of years ago)

The land defines a primary yet subordinate element. Prague was once an underwater territory of volcanos, where receding waters carved the form of the emerging layers of the land, forming valleys and hills. The residues of this ancient background continue to influence the arrangement and the possibility of its settlements.

The People (5000 years ago)

People constitute an important element in the settlement of the land. From Neolithic gatherings on hilltops, and prehistoric roads down to the founding Slavic tribes, temporary clusters accumulated into the rock formations and the terrain, in assemblies of wood and stone. Connected with forests, hunting, gathering and living from the land, the people linked their multiple centres of trading, over time, along these natural topological lines.

The Markets (800)

The markets act as a socio-economic element, organising the people of the land. The topography pre-determined the locations of markets and settlements and consequently the topology of social hierarchies and positions of power in the landscape. One of the largest and busiest places for trading was at the flat plains along the bend of the Vltava River, the centre of which later developed as the Old Town Square.

The Law (870)

The law establishes the element of order as well as the security of society on the land. Markets, gates and towers, sacred temples and houses of stone and wood were protected and defined by town walls, an institution of power and community. Princes, emperors and kings ruled from the Castle. They represented collective security through their laws, unifying and protecting the land, the people and the markets.

The Sacred (973)

The Sacred embodies the higher meaning of the land. The Holy Orders of the city have played a transformative role within the establishment of the built landscape. From the founding of the Bishopric of Prague in 973, the sacred evolved through chapels, churches, cloisters, monasteries and fortresses but always under a larger European order. It is a symbolic law that influences the relation between the individual and its world.

The Hybrid (1348)

The hybrid, of the geological, the social and the sacred, forms the fabric of the city's space. The topological, historical and cosmological — together as one — become the embedded infrastructure of the city, its multi-layered root. In the space between terrain and tower — the vertical space framing both ground and symbol — the story of the city comes to life.

罗马风时期

?—870—?

对于各种历史风格出现的确切日期，不同的地区有不同的定义，很多是尽可能精确的估计。也许罗马风始于公元前27年（独立罗马拱门出现），又或许始于1002年波希米亚被纳入神圣罗马帝国时。870年，随着布拉格城堡的建立，罗马风（其在1818年以后才被认知，考古学家德热维尔首先称这种建筑为"罗马风"）能够在波希米亚被辨认出来。这个带有教诲性的时代富有象征和仪式感。君臣之间的等级制度反映了神权与土地之间的关系，这种礼仪形制被输出给标志性、清修的、永久的建筑中，并与临时定居点形成对比。

布拉格是一个丘陵起伏和山谷盘桓的地区，既不平坦也不呈网格状，而是拓扑和非线性的。它保留了地质变迁的历史印记。石头建筑的出现标志着这片土地上开始建造城市。地景中的场所设定反映出不同的阶层在此居住：统治者定居在有天然防护的前哨，教会被安置在"圣丘"，普通民众则在起伏的地景中安家，大地上的教堂、修道院、城堡和住宅，再次强调了天地间明确的二元等级划分。

在此定居的有斯拉夫人、日耳曼人、意大利人和犹太人等。其中有许多说德语的农夫、商人和工匠来自帝国北部和西部地区，包括基督徒和犹太人，在12世纪晚期到13世纪迁移到此，并且带来了他们的技艺与生活方式。

此时，市场是一个由管理者和经营者（后来发展为商人、企业主、银行家等）组成的日常步行交易网络空间，市场也成为人民与统治者之间交流的媒介。在市场内，法律的力量被发散、分配并增长。

统治者通过修建城墙围合出自己的领地，其中包括人们的生活区域、城市的贸易中心以及教堂。皇室通过联姻等来维持社会和平和稳定，皇室贵族和平民百姓之间界限分明。当波希米亚王权在11世纪初期并入神圣罗马帝国时，继承了源于罗马帝国法的法律秩序。在社会其他的中坚力量中，骑士和神职人员群体是教会与皇室的结合，服务于最高权力。教会从意大利带来高度集中的财富与权力，以及许多新的艺术表现形式。彼时，教会是知识的守护者、文化的生产者，代表着最高真理权威。修道院被视为艺术与语言的保护者，是天地间绝对稳定的中心。

罗马风时期的布拉格
The Romanesque City

教会的官方语言是拉丁语，在当时是一种不被大多数普通民众理解的语言。罗马风建筑的出现为那个文盲占绝大多数的社会提供了可供理解的图像符号，它将预先设定好的内容通过人们熟悉的形式传递给民众。虽然当时还没有人文主义的概念，但是人的内心体验已经被描绘在有关苦难与救赎的艺术作品之中。那时人文主义还不是一个可供讨论的主题，并不成为一种公共形式。

在"人文主义"和"个体"这些概念出现之前，事物并非以其自身独立的形式存在于这个世界中，例如圣像、圣物、永恒、真理，这些意象都是以其原型来表现。这也就是为什么扁平的圣人金像并不是为了表现真实感，而是以象征的方式去展现一种用肖像来表达的严格秩序，与预先的设定只有极其细微的差别。

坚厚的石墙
Solid Stone Walls

圆形的拱券
Round Arches

狭窄的窗户
Narrow Windows

叙事壁画
Illustrative Murals

　　此时，观念与技术的局限也反映在建筑中：人们无法使用更为复杂的空间形式来进行建造，因此，建筑形式简单、布局紧凑——圆形的拱顶、厚实的墙体、紧密石砌中封闭狭窄的通道、细长的采光口，所有这些建筑元素都体现了这一时期厚重稳固的建筑风格。

　　以下是罗马风时期的建筑代表。

1. 高堡的圣马丁圆形小教堂
St. Martin's Rotunda at
Vyšehrad Castle

THE ROMANESQUE

?–870–?

For the exact dates of this or any historical period, the range varies in different regions, and is probably an estimation at best. Perhaps the Romanesque begins before the year 27 BC (with the free-standing Roman arch), or perhaps in 1002 when the Bohemian lands were integrated into the Holy Roman Empire. With the founding of Prague Castle, in 870, the Romanesque (as it has come to be known only after 1818 when archaeologist Charles de Gerville first called the architecture "Romane") can be identified in Bohemia. It was a time of didactic character, of instructive rituals and symbols. Instilled in a feudal architecture of iconic, monastic, permanent forms, and contrasted with temporary settlements, the hierarchy established between princes and people reflected that between the sacred and the earth.

The city landscape, a region of hills and winding valleys is neither flat nor gridded, but topological and non-linear. It retains an imprint of the dynamic fluidity of its geological history. The beginnings of stone architecture mark the foundation of this urban territory. A place in this landscape indicated a social position within it. This equivalence signalled the merging of two realms: that of the land (topological), and that of the social order (symbolic).

Temples, cloisters, castles and homesteads on the land, reiterated a clear hierarchy in the duality of heaven and earth. The royal order settled on naturally fortified outposts, the ecclesiastical order on "sacred mounds", with the people living within the curves and creases of the landscape, each amongst their own relations and affairs.

The people, among whom were Slavic, Germanic, Jewish, and Latin groups settled the land. Many German-speaking farmers, traders, and craftsmen from the north and from western parts of the Empire, both Christians and Jews, moved into these areas by the late 12th–13th centuries, and brought with them their crafts and lifestyles.

The markets, a network of spaces of exchange along the natural paths of the land, composed of managers and operators (later to become guilds, merchants, bankers, industrialists, corporations, and…), became established as an intermediary between the people and the Crown. The markets established a space for visitors and a place of prestige amongst the people. Within the frame of the markets, legal power was decentralised, distributed and growing.

The law of the Crown offered a form of protection: a wall. The wall established a place of governance which included the activities of the territory, the trading centres and the churches in the city. Maintaining peace and social stability, the Crown developed the authority to protect the land and the people, through laws dealing with marriage and competing lines of heredity. Royal blood and royal households maintained society along absolutely clear boundaries, within every strata, according to established norms.

The sacred, when merged with the Holy Roman Empire in the early 11th century, inherited a legal order deriving from Roman Imperial law. Among the other intermediary pillars of society, Knights and Holy Orders, hybrids of Church and Crown, would serve the interests of the higher powers. The Church also brought from Italy a high degree of concentrated wealth and power, as well as many forms of art and new representations. The Church was the guardian of knowledge, the producer of culture, and a representative of the absolute authority on the highest truth of the time. Monasteries were the caretakers of art and language; the centres of a cosmic universe in an absolutely defined and stable world.

The language of the Church was Latin, a language that most people could not understand. The Romanesque was an era of icons offered to a predominantly illiterate society. Pre-established meanings were translated to the people through familiar forms. While there was no culture of humanism, there was a visceral experience of the human body, depicted in an art of suffering and salvation. Humanism was not yet a debated theme, it had not yet a public form.

Before the emergence of Humanism and "the individual" — as a figure of its own form in the world — the iconic, the sacred, the permanent, and the true, were represented in prototypes. Flattened figures of gold-painted saints were not intended to be realistic, but to illustrate an iconographic display of a strict order via symbols, with a minimal degree of nuance beyond the pre-established codes.

Meaning, during this historical time, was communicated with a limited formal vocabulary. This restriction was also reflected in the architecture and in a construction technology that precluded the use of more complex spatial forms. Simple, centralised, round vaults with thick, solid walls; confined and narrow entries through compact and dense layers of stone; thin, narrow cuts for light — were the expression of this heavy and immutable style.

The following example can be considered as both an icon and an emblem of the Romanesque time.

高堡的圣马丁圆形小教堂
St. Martin's Rotunda at Vyšehrad Castle

地址：V Pevnosti, 128 00 Praha 2, Vyšehrad
交通：metro C Stop Vyšehrad（步行约 10 分钟）
建造时间：约 1070 年

☐ 罗马风／ROMANESQUE
■ 新罗马风／NEO-ROMANESQUE

圣马丁圆形小教堂位于维舍赫拉德（高堡），在 1070 年左右作为弗拉季斯拉夫二世的私人教堂建造。该教堂是布拉格现存最古老的建筑，也是保存最完好的罗马风建筑。

维舍赫拉德（高堡）被誉为布拉格的诞生地，它的历史可以追溯到莉布斯公主时期，她是普舍美斯王的妻子，被尊为这座城市的母亲。普舍美斯王朝的早期统治者就定居并建城堡在此。直到它在三十年战争（1618 年开始）期间被破坏之前，维舍赫拉德的皇家城堡一直与赫拉德坎尼的布拉格城堡互为镜像。维舍赫拉德后来用于军事防卫，并在 19 世纪时重建其防御工事时才有了今天的外观。

圣马丁的圆形小教堂采用轴向式和中心式相结合的建筑形式，这一建筑形式源于早期的摩拉维亚圆形建筑类型(此时圣维特大教堂也被修造成这种样式)。西向轴线的建筑入口通往东端面向耶路撒冷圣地的祭坛。教堂的后殿向东延伸出一个线性的集中空间，可以沉思祷告和观瞻教义。

从平面和剖面布局中看，这是一个带有小窗和厚墙的圆拱建筑。教堂主空间的直径为 6.5 米，并附着一个直径为 2.2 米的半圆形后殿。建筑为小方石砌筑，现在墙体中还能看到最原始的砖石。这些小而密集的石块围合成光滑的弧面，使其墙面达到完美。一个浅抛物线的穹顶横跨空间制高点，另一个穹顶修造在祭坛之上。教堂的圆顶上方还矗立着一个石灯笼，由石、木和金属制作而成。

教堂的墙壁厚约 1 米。通过冰冷石墙上坚实的木门进入内部，是一个开敞、无梁的高耸空间。轻盈的彩绘圆拱顶抬升并照亮了整个空间，使人感觉安全、避世而静谧，形成一方由几何图形构成的精神净土。

室内罗马风时期的艺术作品和其象征意义也独立于外部世界。例如标志性的金叶和彩色图像描绘出一些隐喻性信

息，而光泽耀眼的图像则呈现出一种扁平化的而且永恒的特征。礼拜的神圣意义因此被描绘成亘古不变的。然而，原始的壁画却没有幸存下来。

随着时代的变迁，教堂也服务于新的功能。1878—1880年，它被修复成新罗马风样式：这是19世纪人们对宗教本源的诠释。教堂如今的外观大部分受这一时期修缮的影响，例如对穹顶底座用铁件包裹加固。穹顶上壁画和细节还采用罗马风时期的材料色调和绘画技巧，但是人物表现则运用新罗马风格：圆润而富有表现力，却没有原来的平整感。

窗户和入口石门也是以修正主义的手法进行重建，摒弃了其简单的原始形式。

1782年后，圆形教堂被关闭；1841年险些因修路被拆除，后来人们将建筑的西入口移到南面它才得以幸存。人们可以从教堂上观察到建筑的各种历史痕迹：1757年普鲁士围攻布拉格留下的一枚炮弹还留在立面上，而砖石之间隐约可见原西入口和后殿上方老窗的痕迹。

在布拉格新城还有另外两座同类的小教堂也被保存下来，分别是圣十字教堂和圣朗吉努斯教堂，这两座教堂的外部环境后来也都发生了改变。

....at the edge of a walled enclave, within nature, itself its own fortification, in the forest outside Vyšehrad Castle.

The Rotunda was built as a private chapel for Prince Vratislav II around 1070, just outside the walls of his Royal Castle. It is the oldest and best preserved example of Romanesque architecture in Prague. The purpose of such a chapel was to form a place of sacred space within the natural landscape; a retreat from the world into a universe within itself. The intimate scale and the independent position give this type of building its sacred character.

The earliest settlements of Vyšehrad are unknown, but can be traced to Princess Libuše, mother of the Přemyšlíd dynasty and of the city, who had the first vision of Prague here in the 8th century. Until its destruction during the 30 Years' War, the Royal Castle at Vyšehrad was a mirror to the Prague Castle on Hradčany. Both were built as small fortresses, due to their geological positions atop steep cliffs overlooking the river and the city. Vyšehrad was subsequently used by the military, and the reconstruction of its fortifications in the 19th century give it the appearance it has today of a walled enclave.

St. Martin's Rotunda uses a combination of axial and centroid forms, derived from the earlier model of the Moravian rotunda (at this time, St. Vitus was built as this type).

The original entry along the west axis directly faced the sacred space of the altar to the east, towards Jerusalem. With its extended eastern apse, both linear and centralised, the chapel was a focused space of contemplation and visual doctrine.

The architecture, in both plan and section, consists of a circular arch with small windows and thick walls. The main space is 6.5 metres in diameter with a 2.2-metre diameter apse defining the altar. It was built with precision and craft, primarily of small ashlar stones, and still has its original masonry. These small and dense stones were used to achieve a smooth and constant curve of the wall. A shallow parabolic dome spans the high point and another dome was built above the altar. The chapel was topped with a stone lantern on its dome and finished with details in hand-carved stone, wood and metal.

The solid walls are nearly 1 metre thick. At the single point of entry through a solid wooden door, passing through a stone cold mass, one perceives an open vertical space — a house without roof beams. A domed and painted ceiling made the space feel safe, illuminated and up-lifting. It is an insulated and silent space; its interior exists as a place outside the world, defined by its internal geometry.

The art of the Romanesque era and its symbolism was also independent from the outside world. In iconic representations of gold leaf and coloured figures, metaphorical narratives were depicted while figures and forms in this illuminated imagery assumed a flattened, almost timeless character. The sacred meanings of liturgical messages were portrayed through them as fundamentally unchanging. These original murals of the Rotunda, however, have not survived.

The chapel served new purposes as the times changed. After 1782, the Rotunda was closed and was to be demolished in 1841 for the construction of a road. Instead, its original western entrance was re-located to the south and the building was saved. Between 1878 and 1880 it was renovated in a Neo-Romanesque style: a 19th century interpretation of the original. The paintings and details aimed to follow the Romanesque material palette and painting technique, but the representation of figures was made to suit 19th century tastes: rounded and expressive, without the original flatness. The windows and the stone entry portal were also reconstructed in this revisionist manner, which departed from the simpler original forms. Much of the appearance today is influenced by these more recent renovations, such as the iron reinforcement which wraps the base of the dome. Other elements of the building's history can be observed: a cannonball, from a Prussian siege of Prague in 1757, remains in the facade. The traces of the former western entry and older windows above the apse, can be discerned between the lines of stone.

There are two chapels of the same type still preserved in the New Town, the Chapel of the Holy Cross and the St. Longinus Chapel, although both have since lost their grounds.

哥特时期

?—1348—?

……也许这个时期起源于1144年法国圣丹尼斯教堂的重修完成，或者开始于1215年的第四次拉特兰宗教会议，也有可能起源于弗里德里希二世在1228年领导的第六次十字军东征。也许在1225年，当圣阿格尼丝公主选择在自己的领地独立时，在老城区内建造哥特式建筑是上层阶级妇女对新权利的诉求。或许在1550年乔治·瓦萨里对"哥特人"批判时才出现了哥特式。当然，最迟不晚于1348年，人们能看到一种先进的哥特秩序正在重建波希米亚。这种重建一直持续到15世纪初扬·胡斯去世后，他在伯利恒教堂宣扬妇女在教会中的平等地位，并为斯拉夫人争取更多的权利，因而于1415年在德国康斯坦茨被处以火刑。

在1230年和1257年，布拉格的老城和小城先后获得自治权，并在皇室的指引下发展，皇室也维持着市场和教会的秩序。手工业和商业行会的政治和经济权力不断提高。在手工业鼎盛时期，新兴人道主义价值观和人权观开启了一定程度的民主政体。到1348年，查理四世扩张了城市，并将其作为神圣罗马帝国和波希米亚王国的首都，标志着神权和民法新的结合，为漫长的黄金统治期奠定了坚实基础。此时，城市的主教堂终于取代修道院成为主导建筑，教堂不再是纯粹的宗教建筑，而成为城市公共生活的中心。市民希望主教堂是美丽、欢乐和生机勃勃的。

12世纪中叶，哥特艺术从法国传入布拉格，可以看到建筑形式从沉重的罗马风样式中解放出来，并逐渐转变为轻盈的哥特式。这一转变还表现在罗马风建筑既定的等级体系中出现了一种早期的人文主义形式，图像化的表现形式逐渐减少，开始明确区分一般性和特殊性的建筑表达。在哥特时期的建筑中结构与形式分离，开始表现出具有独立性的个体元素。14世纪，查理四世将才华横溢的法国建筑师阿拉斯的马蒂亚斯和德国建筑师彼得·帕勒请到布拉格皇宫，参与圣维特大教堂的设计建造，标志着布拉格建筑进入哥特时代。

在这一时期，哥特式对独立个体发展的强调远超过罗马风装饰的象征意义和表面品质。哥特式对人类个体自主性的认识，与脱离厚实墙体而存在的独立

哥特时期的布拉格
The Gothic City

人像相呼应。这与当时不断扩张的市场、分权的社会,以及教会及其资助者在新的地方化秩序中分权并行。

哥特式大教堂采用骨架券作为拱顶的承重构件,并采用尖券和尖拱,打开了结构和形式之间的空间,减轻了建筑的重量并获得更多光照。在人类的定居点转变为城镇时,这种建造方法可以更快地实现,但大教堂可能需要几代人才能建造完成。作为装饰和赞美城市的纪念碑,哥特式建筑外部更多体现着市民阶层对现实生活的热爱,展示着工匠卓越的技艺和城市的富足与独立。

波希米亚的哥特式建筑在查理四世统治时期达到高潮。

早期的尖拱券
Early Pointed Arch

晚期的尖拱券
Late Pointed Arch

查理四世与新城

查理四世（1316—1378）是波希米亚国王及神圣罗马帝国皇帝。在其统治时期，波希米亚成为神圣罗马帝国的核心，布拉格成为神圣罗马帝国的首都，这是中世纪捷克最强盛的时期。1348年3月，查理四世下令以今天的查理广场为中心，修筑新城。兴建新城是为了解决布拉格面临的空间和环境问题，为城市发展提供足够的空间，满足其成为大都市的各种要求。

新城由彼得·帕勒具体规划，面积相当于老城的三倍。新的道路、商贸点、标志建筑、基础设施、大学（中欧最古老的大学）、教堂等在新城中应运而生。新城的扩张使布拉格在1372年拥有了4万居民，成为阿尔卑斯山以北的第四大城市，布拉格石头城市的形象得以展现。由于吸引了本土的商人和工匠，新城成为布拉格的繁华区域，并发展出自己（捷克）强大的贵族阶级。地区经济贸易的自主和繁荣使得布拉格成为当时欧洲最富庶的城市之一。新城规划成为欧洲城市从哥特到文艺复兴转换的典型，历史和社会身份在城市空间内共生并产生新的含义。

以下的例子是在老城和城堡区的建设项目，它们在哥特早期和晚期得以兴建。

2. 犹太区老 - 新犹太教堂
Old-New Synagogue of the Jewish Town

3. 布拉格城堡内的圣维特大教堂
St. Vitus Cathedral in Prague Castle

4. 查理大桥
Charles Bridge

5. 老城广场的旧市政厅
Old Town Hall

THE GOTHIC

?–1348–?

... perhaps it began in 1144 in Saint-Denis or in 1215 with the Fourth Council of the Lateran. Or perhaps when Emperor Frederick II led the Sixth Crusade in 1228. Perhaps in 1225 when Princess Saint Agnes chose independence within her own compound, still standing within Old Town, as claims of the new rights for upper class women. Or perhaps only in 1550 with Giorgio Vasari's critique of the "Goths". Certainly by 1348, an advanced Gothic order was re-establishing Bohemia. It continued through the early 15th century after Jan Hus - who preached at Betlehem Chapel for equal rights for women in the Church and increased rights for the Slavs - was burnt in Konstanz, Germany for heresy in 1415.

Changes in society began to emerge. Revisions to the legislation of the political order, and a shift towards the decentralisation of power and to social compassion began a transfer of autonomy to local authorities.

The people: The accumulation of contradictions and internal tensions within the absolute established laws, over time, produced political pressure for change. Demands for local governance, for the rights of women and for individualism gradually developed.

The markets: Rising political and economic power of the guilds and trades, and emerging Humanist values and legal rights for the people and the merchant families, initiated a new period of civil independence under the Crown. The founding of Old Town in 1230 and Lesser Town in 1257 and the Jewish Town in 1262, as self-governing authorities, granted these towns newly established powers.

The law: The newly formed towns would develop under the guidance of the Crown, which would maintain order over both the markets and the Church. By 1348, Charles IV establishedthe foundations for a long golden reign by marking a new union of sacred and civil law as the capital of both the Holy Roman Empire and the Kingdom of Bohemia.

The sacred: Under the growing sponsorship of noble families, every church could now offer something unique. A cult of pilgrimage and the processional experience developed along with significant changes in the Arts. Temples dedicated to Mary — as woman and mother — became central themes, and representation shifted emphasis away from suffering and pain towards compassion and love, as Jesus became represented as an infant.

In architecture, we see a gradual release from the heavy Romanesque to the uplifting of the Gothic. From the established hierarchies of the Romanesque emerges an early form of Humanism, with increasingly less iconographic forms. These began to articulate a distinction between the generic and the particular. In the architecture of the Gothic time, the separation of structure and form began to express an emerging independence of individual elements.

Throughout the period, an emphasis on the individual developed beyond the symbolism of the Romanesque ornamentation and its surface qualities. The recognition of individual human autonomy paralleled the gradual emergence in "the architectural wall" of the free-standing figure as distinct from the solid mass. This took place in parallel to the growing market and the decentralisation of society within new localised affairs of the churches and their civic patrons.

Gothic arches, ribbed vaults and buttresses allowed for the abandonment of mass for light, through the opening of a space between structure and form. Cathedrals could take generations to construct, so units and patterns, (such as Roman numerals) offered an instructive architecture as well as the formulaic flexibility inherent in them. The new architectural expression, the lancet style, gave form to previously unseen structural forces through its use of pointed arches, engineering advances and exposed structural elements. This coincided with accelerated construction in general as settlements became towns.

Charles IV and the New Town

Founded in 1348 during the reign of Holy Roman Emperor and King of Bohemia, Charles IV (1316–1378), the New Town was the largest construction in Europe after Rome. Prague was to become a New Jerusalem, an economic and spiritual melting pot. The establishment of a fixed layout, of a particular manner and order of rules — for the university, monasteries, markets, roads and walls — sustained the New Town with cultural and civic infrastructure. As a class of merchants (living in stone houses) became established, the stone figure of the New Town developed its own autonomy in parallel. Historical and social identities could acquire new meanings within the space of the New Town, through the logic of growing together.

Roads, trade quarters, landmarks, squares and grounds, universities, churches and other institutions emerged. Towers, temples of stone and naturally contoured streets were established within the formation of the

早期的扶壁
Early Buttress

晚期的飞扶壁
Late Buttress

尖头窗
Lancet Style Windows

尖拱顶
Lancet Style Vaults

city, and in time became its solid foundations. The markets became centres of civic life. The guilds, crafts and unions, as well as merchants and others found incentive to accumulate the growth of their New Town, in the spaces of exchange between the new markets. The trade activities which developed produced a network of socially and topologically specific identities, (for example, the cow market was a pasture with the meat markets nearby; the horse market connected two city wall gates with the hay market nearby, etc.). These topological, functional and economic factors formed the basis of a social townscape — and today's neighbourhoods. Economic autonomy, security, and mutual dependence allowed the trades and towns to flourish during European crises (there was no Bubonic plague).

The following examples can be considered the founding centre-points of the Old Town, re-established during the early and late stages of the Gothic time.

犹太区老 - 新犹太教堂
Old-New Synagogue of the Jewish Town

地址：Maiselova 18, 110 01 Praha 1, Josefov
交通：Tram 2, 14, 17, 18, 93 Stop Staroměstská（5分钟步行）
建造时间：约 1270 年

☐ 哥特／GOTHIC
☐ 文艺复兴／RENAISSANCE
☐ 巴洛克／BAROQUE
☐ 新古典主义／NEO-CLASSICAL

老-新犹太教堂的历史最早可追溯至1270年，它是欧洲现存最古老的犹太教堂，早先被称作新犹太教堂或大犹太教堂，直到16世纪后期其他犹太教堂相继出现后，才采用老-新犹太教堂的名称。教堂位于布拉格犹太人聚居区（下文简称犹太区）的中心地带。犹太区最早的地势向伏尔塔瓦河倾斜，因此老-新犹太教堂在建时矗立在区域的高处，但随着时间变迁，今天它却处在城市低处。

由于它最初的建成环境信息大多已经消失，因此必须将其放在历史语境中才能被重新想象。老城内独立的犹太人定居点可追溯至10世纪，并且在13世纪早期获得官方认可，犹太人获准在此生活、旅行和贸易，犹太区成为了一个功能完善的城镇，内部有住宅、市场、教堂、商肆、学校以及沿河的活动场地等。但犹太人不被允许住在犹太区以外的地方，也不可以拆除现有的犹太区内的建筑物。因此，几个世纪以来人们只能在已有房屋的基础上进行改建。

1867年奥匈帝国法律的修改赋予犹太人在帝国内部平等以及可以在任何地方居住的权利，这标志着犹太区的结束，也为社区的重新发展以及融入现代城市创造了契机。为抵御频繁泛滥的洪水，

河堤最终于19世纪末被建造起来。同期，布拉格模仿巴黎进行城市改造，因犹太区内拥挤且杂乱无章的建筑不符合布拉格城市建设新的标准，因此它们从1893年开始被拆除。1895—1910年在此区域大规模为城市新兴的中上阶层建设了住宅，这些住宅大多为那个时代最流行的新艺术风格，还建造了切赫桥和巴黎大街。这也使得布拉格成为现存新艺术运动风格建筑最多的城市之一。

只有少数犹太教堂及犹太人墓地、犹太市政厅、梅塞洛瓦大街和墓地周边地区在改造中被保留下来，可以从中看到犹太区的过去。在20世纪40年代，老-新犹太教堂险些被再次拆除，因为希特

勒打算把犹太区改造成"种族灭绝"博物馆，但最终它幸免于难。如今，这座犹太教堂已经成为珍贵的文化遗产，并且是布拉格最受游客欢迎的景点之一。

它之所以如此不同寻常，是因为它抽身于被摧毁的周围环境，但却没有遗失任何信息。站在老-新犹太教堂面前，立于犹太市政厅和犹太人墓地之间，就可以认出原本这里是老城的低地。老-新犹太教堂矗立在两条街道的交会处，这里是唯一的一条原犹太街区留下来的道路。随着现代城市的建立，除了梅塞洛瓦大街之外，所有原来街道的痕迹都消失了。老-新犹太教堂像是脱离于周边时空而独存，它原先所属的历史及文脉都已经随风而逝。

教堂外观独特：4个不同的立面形成4种不同的特征，建筑庞大的主体四周被小体量的附属建筑围绕支撑着，这些小体量建筑及入口是在14世纪至18世纪的不同时期增添的。由于一度处于一个更密集紧凑的环境中，这栋建筑似乎没有清晰的入口，只有一扇小门进入，但里面是矗立在原始地坪上的最原始的大门，也是布拉格最古老并保存完好的石刻大门之一。

老-新犹太教堂外观特色是裸露砖墙、大坡屋顶、小窗户以及6个扁方形嵌入在墙和屋顶中的扶壁。建筑屋顶是独立结构，只有通过梯子从东立面上的一个小开口才能到达。有些人通过回忆戈勒姆（泥人）的神话故事来解释建造如此高大屋顶的目的。据说，戈勒姆是为了保护犹太区及其居民而创建的，它就住在这个屋顶内。当然，屋顶空间是空的（戈勒姆可能就是建筑本身）。从事教堂修建的皇家石匠，还在不远的河边建造了圣阿格尼丝修道院。

建筑内部哥特式拱顶有9米高，但从建筑外观并不能看出这种结构形式。两个巨大的八角形石柱支撑着6个砖拱顶，并构成了一个独立、高耸的塔状空间：紧凑而私密，适合小型集会和其他宗教活动。建筑内部空间面东，朝向耶路撒冷。19世纪末重建时，屋顶得到修缮，尽管当时建筑被"净化"，已经风化暴露的砖块掩盖在灰泥里，但内部仍然显露原来的建筑布局，并保存了它巴洛克时期的家具。

弗兰兹·卡夫卡的一生跨越了犹太区拆除的前后。居住此地的他体会到布拉格从一种城市形式到另一种城市形式的转变。因此，卡夫卡的《变形记》可以从这一历史语境来解读。

... in the heart of the Jewish Town, its original terrain now 2 metres below the ground level of the city and its original context mostly gone.

Unlike the churches and cathedrals of the Gothic period, which aimed to communicate and symbolise a message for the society, the Gothic architecture of the Old-New Synagogue was not designed to be a representative object from the outside. Its priority was on its interior. The building is, essentially, a single room. Its ambition was to be a multi-purpose civic house, used for assembly and for study.

The Old-New Synagogue must be re-imagined in terms of its historical context, now mostly demolished. Within the Old Town walls was an independent Jewish settlement of Prague, dating from the 10th century, and granted official status as a legal quarter in1262. Jews were permitted to live, travel and trade under the higher laws of the land. The Jewish Town was a fully functional town with clusters of houses, markets, temples, businesses, schools and activities along the river, but operating within Imperial law administrated by the Crown. Not allowed to live elsewhere or to demolish existing buildings, the people built and renovated their houses upon themselves, within this area, over centuries.

Changes to the Imperial law in 1867 granted the Jewish people equality within the Empire and the right to live anywhere. This marked the end of the Jewish Town. This also opened the neighbourhood for redevelopment and its integration into a modernising city. Due to frequent floods, the river embankment was eventually built at the end of the 19th century. The haphazard structures of the Jewish Town did not meet new 19th century standards and their demolition was to begin in 1893. The residential development project from 1895 to 1910 was the first building development project for mass housing for a new upper middle class market in the centre of the city. It included the construction of the Čechův bridge and Pařížská boulevard.

Only the few synagogues, the cemetery, the Jewish Town Hall, Maiselová Street and the land around the cemetery remain from the Jewish Town of the past. The Old-New Synagogue was also spared from demolition a second time in the 1940s, as Hitler intended to preserve the Jewish Town as a "museum of an extinct race" . Fortunately, this did not happen, and it has become an exceptional object of cultural admiration today, one of the most popular sites for visitors of Prague. It is still functioning as a synagogue.

The oldest synagogue remaining in central Europe, dating from 1270 was once the newest synagogue in Prague, and eventually became the Old-New Synagogue as more were built. The Jewish Town once also sloped towards the river and the synagogue, built upon the original ground, once stood at a higher point of the landscape but stands 2 metres below ground level today.

It appears so unusual because it is disentangled from its surroundings which have been demolished — but it is not lost, (in fact, it contains its front door within itself). The synagogue stands at an intersection of two streets, the only paths remaining from the original Jewish Town. When one stands in front of the synagogue it is possible to recognise the lower ground of the older city, between the Jewish Town Hall and the Jewish cemetery. As the late 19th century city was built around it, all traces of the ground

disappeared apart from the topography of Maiselová Street. Its place within the historical and cultural fabric has been erased in space, and the Old-New synagogue finds itself standing on a separate ground of time, its origin internalised.

The elements of its structure are very particular and have no remaining precedent, with 4 different sides having 4 different characters. The massive and compact form is surrounded at ground level by smaller volumes which support this main space and have separate entrances. They were added at various times from the 14th to the 18th century. Once in a more dense environment, the building appears to have no clear front, only a small door, but inside there is the original door upon the original ground — one of the oldest and best preserved carved stone portals in Prague.

Royal stonemasons who worked on the construction also built the Convent of Saint Agnes not far down the river. Notable elements of its exterior are the brickwork, small windows, the six, flat and square buttresses embedded in the walls and the massive roof. The roof is closed with a pair of tall, exposed brick walls. It is a separate structure, accessible only from a ladder through a small opening on the eastern facade. Some explain the purpose for such a large and tall roof by recalling the mythical

figure of the Golem. The Golem, said to have been created to protect the Jewish Town and its people, is also said to be living inside this roof. The roof space however, is empty. (The Golem may actually be the building itself…)

The Gothic vaulted interior is 9 metres high, and its expression is not revealed in the architectural exterior apart from the understated buttresses. Inside, two large, octagonal columns of stone support 6 brick vaults which define the single room. The towering space is quite compact and intimate, suitable for its smaller gatherings and its domestic-sacred character. The interior faces east towards Jerusalem and in the late 19th century, was renovated as the roof was reconstructed. Despite being "cleaned up" at this time, with the exposed and aged bricks hidden in stucco, there is still the original building composition and its original Baroque furniture inside.

Franz Kafka's life spanned the duration of the Jewish Town's demolition. While it was his neighbourhood, he lived within the transformation from the one city form to the other. Kafka's *Metamorphosis* could be read in light of this historical context.

布拉格城堡内的圣维特大教堂
St. Vitus Cathedral in Prague Castle

地址：III. nádvoří 48/2, 119 01 Praha 1, Hradčany
交通：Tram 5, 7, 11, 12, 15, 20, 22, 23, 41, 97 **Stop** Malostranské náměstí
（10 分钟步行）
Tram 23, 41 **Stop** Pražský hrad（5 分钟步行）
建造时间：1344—1929 年
建筑师：阿拉斯的马蒂亚斯（Matthias of Arras）、彼得·帕勒（Peter Parler）

- 罗马风 / ROMANESQUE
- 哥特 / GOTHIC
- 文艺复兴 / RENAISSANCE
- 巴洛克 / BAROQUE
- 新哥特 / NEO-GOTHIC
- 现代主义 / MODERNISM
- 当代 / CONTEMPORARY

从伏尔塔瓦河对岸望去，雄伟的布拉格城堡矗立在赫拉德坎尼山上，成为布拉格最重要的视觉图像。布拉格城堡是捷克历代君王的住所，也是当今国家元首的府邸，建筑面积达 7.5 万平方米，是欧洲最大的城堡。如同一个微缩城市，布拉格的许多变迁都可以从城堡上寻找到线索。布拉格城堡内的圣维特大教堂是波希米亚地区最大和最重要的教堂，也是这片土地上永恒的象征。

圣维特大教堂所处位置最早是一座 9 世纪的罗马圆厅小教堂，随后从 1060 年起扩建为罗马三殿式教堂。哥特式大教堂是波希米亚国王和后来的神圣罗马帝国皇帝查理四世下令从 1344 年开始在原基础上，聘请法国人阿拉斯的马蒂亚斯和德国人彼得·帕勒先后负责修建。大教堂不仅被作为国王和皇帝的朝觐地点和葬礼墓地，而且还被视作查理四世统治时期教会与皇室联合的象征。

大教堂分三个部分：哥特式的东半部、带文艺复兴顶部和巴洛克圆顶的哥特钟楼和新哥特式的西半部。各部分之间的连接清晰显示在教堂内的翼廊与中殿的交叉处。东半部分哥特式后殿始建于 14 世纪中叶，有着结构清晰的独立双飞扶壁，体现了哥特建筑的最高表现形式。这些飞扶壁从外部抵御屋顶的重力作用，减轻墙体的负荷并释放出内部空间，形态比例短而圆，保留了早期圆形穹窿大厅的样式。黄金大门，是原始大教堂入口，位于南侧。华美的三拱哥特式门廊由彼得·帕勒设计，上方装饰有威尼斯工匠制作的金色马赛克镶嵌画——《最后的审判》，在阳光下熠熠生辉，成为天与地、阳光与大地、物质与灵魂间的三座桥梁。国王正是通过这扇大门，进入主教堂举行加冕典礼。

教堂钟楼在 15 世纪后半叶加入了文艺复兴风格的顶部和巴洛克的塔尖；西半部的正殿、屋顶、大门以及新哥特式的塔楼在 1867 年至 1928 年间建造完成。中殿北窗上有捷克著名艺术家阿尔丰斯·穆夏在 20 世纪初创作的彩绘玻璃画。教堂的修建历经几乎 600 年，一直延续到 20 世纪才全部完成，是由不同时代的多名建筑师先后参与共同完成的。

圣维特大教堂对后期哥特式建筑风格在中欧地区的发展产生了巨大的影响。特别著名的实例包括维也纳的斯蒂芬大

教堂、斯特拉斯堡主教座堂,以及捷克库特纳霍拉的圣巴巴拉教堂等。

布拉格城堡与广场

从9世纪末的普舍美斯家族的统治开始,经过查理四世和鲁道夫二世、玛利亚·特蕾莎女王,直到托马斯·加里格·马萨里克和瓦茨拉夫·哈维尔总统时代,布拉格城堡及其内外不断进行扩建,其场地构成特征已经完全改变了。

城堡俯瞰整座城市,与伏尔塔瓦河对面的高堡相呼应。在罗马时期,它最初是作为有设防的皇室住所,维护统治、抵御入侵并防止火灾。城堡在哥特时期经过多次重建,随着时间的推移,从一座中部核心、东西两翼的三段式城堡,逐渐扩展为现如今的形态,以适应皇室的发展。圣维特大教堂周围是宫殿、教堂、贵族府邸等。黄金巷中聚集着商人、金匠和艺术家。

几个世纪以来,城堡内部不断进行着修缮和改建。波希米亚首座哥特式-文艺复兴风格的宫殿:弗拉迪斯拉夫大厅始建于1493年。砖石结构的外立面、哥特式的墙体、文艺复兴式的窗户、新古典主义的灰泥装饰、巴洛克风格的柱廊和一些现代的细节,这些风格逐渐随着时间混合在一起。皇权在1621年后的300年里被转移到维也纳,在此期间,修缮主要集中在城堡的内部空间以及外观的简洁庄重上,使其从城堡转变成宫殿,这种新古典主义风格下进行的建造活动一直延续至19世纪。在此期间,防御工事被简化到了最少。这一历史时期中,建筑的特性多表现为理性以及平衡,许多哥特风格中不规则以及不对称的样式都被掩盖了,但仍然可以在宫殿南面的灰泥外墙下看到原始防御城墙的巨大基墩。

1918年,捷克斯洛伐克第一共和国成立后,布拉格城堡的修缮及改建工程主要包括庭院、方尖碑、旗杆、总统府、花园、温室、西班牙宫北门等,由第一任捷克斯洛伐克总统托马斯·加里格·马萨里克委任斯洛文尼亚建筑师约热·普列赤涅克,在1920年至1929年间完成。20世纪90年代到21世纪,布拉格城堡及圣维特大教堂也有修缮及新的设计添加。

... on the tip of Hradčany upon a natural jetty of rock with cliffs on both sides, a counterpoint to Vyšehrad, the Cathedral soars in the sunset above the city.

St. Vitus Cathedral developed through multiple changes over time. At once a synthesis of the whole social fabric and the many transformations which have accumulated within its walls and lands, the Castle is also a microcosm of the city. The continuous evolution and its architectures are echoed in both. Prague Castle, with over 75 000 square metres of rooms, is the largest castle complex in Europe. As the symbolic Crown of the people, and as a sacred tomb of its leaders, the St. Vitus Cathedral in Prague Castle, the most important of Bohemia's cathedrals, is a continuously enduring symbol of the heart of the land.

The first churches of St. Vitus were a four-apse Romanesque chapel from the 9th century, followed by a three-nave Romanesque church from 1060. The 1344 construction of the gothic cathedral was built over these foundations by Matthias of Arras and Peter Parler. It was commissioned not only as a place of pilgrimage and a burial space for kings and emperors, but also as a symbol of the union of the Church and the Crown at the time of the reign of Charles IV. The Golden Gate (the original Cathedral entrance) is the locus of the city of Prague and the meaning of its land. It is the sacred form of the threshold. Seen from the city below, the civic tower represents this synthesis of the land and the Sacred, in a leading display of symbolic authority.

There are notably three parts to the Cathedral: the Gothic structure of the eastern half, the Gothic belfry tower with its Renaissance top and Baroque cupola, and the

Neo-Gothic western half. The connection between the parts is clearly seen within the Cathedral, at the intersection of the transept and the nave. The original entry was from the south, between the tomb of Wenceslas and the tower of the Holy Kingdom. The Golden Gate, with its south facing Gothic mural in golden embroidery, stands in a reflection of the sun, as the three-portal bridge between heaven and earth, light and land, spirit and matter.

The eastern apse was built in the mid-14th century, in the highest form of Gothic, with

its ornately articulated, free-standing double flying buttresses. The apse is of short and round proportions which maintains a trace of the earlier rotunda form. The buttresses support the weight of the roof from outside, lightening the mass of the wall and freeing the space of the interior to be illuminated by coloured glass. While the top of the tower remained unfinished for over 200 years thereafter, it was finally completed with a Baroque top. Over 300 years later, between 1867 and 1928, the Cathedral's western nave and roof, its portal and new Neo-Gothic towers were completed.

Prague Castle and Grounds

From the rule of the Přemyslid family at the end of the 9th century, through the reigns and Golden Ages of Emperors and Kings, Charles IV and Rudolf II, to Queen Maria Theresa, to the presidencies of Tomáš Garrigue Masaryk and Václav Havel, the composition of the Prague Castle and its grounds have been transformed in character both inside and out.

In the Romanesque period, it was originally made as the fortified residence of the Royal family. The Castle overlooked the city and communicated with the Royal residence at Vyšehrad. It was built to withstand fires and invasions and to protect the rulers. Through multiple reconstructions during the Gothic period, from a fortified 3-part castle with its central core, east and west wings, it was built up over time to support the growth of the noble families.

The grounds gradually expanded to include merchants and alchemists and scientists along the Golden Lane. Around St. Vitus Cathedral are palaces, churches, noble households, and halls of governance. There are sculptures, portals and stairs in gardens within the courtyards and along the southern and northern slopes.

Within the Castle walls, interiors were redesigned and new details installed over centuries with meticulous construction in a range of techniques. The first mixed-style Gothic-Renaissance palace in Bohemia, Vladislav Hall, was built in 1493. Among facades of brick and stone, Renaissance windows within Gothic walls and Neo-Classical stucco, Baroque colonnades and modern details — all styles have mixed there over time. In 1621, the Crown moved its seat of power to Vienna for the following 300 years. During this period, renovations focused on the interiors and the simplified and dignified external appearance of the Castle, turning it into a palace. Building works in the Neo-Classical styles continued through the 19th century. During this time, the fortifications were minimised. In a display of rationality and balance, characteristic of this historical time, many of the Gothic asymmetries and irregularities were covered. Traces of the massive piers of the original fortified wall however can still be seen bulging from beneath the stuccoed facade on the southern front of the palace.

Major construction works on both St. Vitus Cathedral and Prague Castle were completed during the Modernist renovation project from 1920 to 1929. New layers were also added to the interiors and gardens in the 1990s, with renovations and small details bringing playful contemporary notes to the historical context.

查理大桥
Charles Bridge

地址：Karlův most, Praha 1
交通：Tram 5, 7, 11, 12, 15, 20, 22, 23, 41, 97 **Stop** Malostranské náměstí
（6 分钟步行）
Tram 2, 14, 17, 18, 93 **Stop** Staroměstská （6 分钟步行）
建造时间：始建于 1357 年
建筑师：彼得·帕勒（Peter Parler）

■ 罗马风 / ROMANESQUE
□ 哥特 / GOTHIC
■ 巴洛克 / BAROQUE
■ 新哥特 / NEO-GOTHIC

查理大桥位于伏尔塔瓦河流经布拉格的城中心，可以说是布拉格的实体和精神象征，也是这座城市在欧洲的缩影。至少从这座桥的地基铺好开始，欧洲的这一半和那一半就一直在互相寻找。东方和西方，同一种文化的两个分支，却代表着不同的传统。

查理大桥始建于1357年，其之前的位置上是一座建于1172年的名为朱迪思的石桥，这座石桥在1342年被洪水摧毁。新的大桥由查理四世委任彼得·帕勒（也是圣维特大教堂的建筑师）于1357年开始建造，最终建成于1390年。直到1841年，这座大桥一直是跨越伏尔塔瓦河的唯一桥梁，也是连接老城、小城到布拉格城堡及邻近地区最重要的通道。这座大桥最初被称为"石桥"或"布拉格桥"，到1870年才被命名为"查理大桥"。

大桥宽约9.5米，有16个桥拱，总长515米，两端有3座桥塔。位于布拉格老城一侧的桥塔修建于14世纪后半叶，不仅是防御工事，也是进入老城的凯旋门，是布拉格晚期哥特建筑最重要的存留之一。位于小城一侧查理大桥尽头的两个桥塔中，稍矮的是12世纪朱迪思桥的残存部分，稍高的建于15世纪中叶，是对老城桥塔的模仿。

大桥采用当地的砂岩建造。由于历史上查理大桥曾遭受过数次浩劫，因此现在查理大桥上只有约30%的石头是最初的。塔楼的外立面细节也已重建，但大桥的高度和外形都是最初的样子。精心设计的钻石形桥墩在水中支撑着桥体巨大的跨度，并受到从水中升起的木结构的保护。它们也曾被用作破冰装置以保护桥梁，并最大限度地减少对桥基的压力。

在三十年战争末期，老城桥塔一侧的桥体在战争中被严重损坏，后来残存的哥特式装饰物也被迫移走。战争结束后，巴洛克式的重建活动反映了权力的变化以及教会在日常生活中新权威的建立。现在桥上的30个人物雕像大多是在1706年到1714年间被放置。这些雕塑在后来被统一用复制品代替，仅有一个为原物，其他原物被移往国家博物馆保存展出。桥上最著名的人物塑像是圣约翰·内波穆克，是桥上最古老的、唯一的铜雕像。传说，如果你反复擦拭雕塑下的铜饰板，你一定会在未来重返布拉格。

在1723年，大桥开始采用油灯照明，1866年，仿哥特式的煤气灯竖立在桥两侧的栏杆上，并沿用至今。1905年，桥面开通了电车后更换为公共汽车。1965—1978年，查理大桥进行了大修，桥墩的稳定性得到了加固，破损石块都得到更换，沥青路面被剔除，此后查理大桥禁止所有车辆交通，变成了一座只允许步行的桥梁。

查理大桥混合了哥特式和巴洛克的样式，组合了皇权和教会的元素，以及中世纪商业集市景观和从城堡眺望的节奏韵律，融入了城市的社交空间，包含了整个布拉格城市精神。如今，查理大桥是一个观看日落、烟火、冬雪和街头演艺的好去处。无论大桥空寂无人或熙熙攘攘，都是它最美丽的时刻。

… the central thread of the city spans the Vltava river and the world.

Charles Bridge is a connection between the scale of the city and the scale of the Kingdom. It links the two sides of the Vltava — through the Old Town and the Lesser Town, to the Royal seats of Prague and Kutna Hora — into the extended landscape. Placed in the centre of the river, it is the critical point along the Royal Way.

For 500 years, it was the sole connecting road between the two towns. While serving also a symbolic purpose, it was intended primarily as a piece of infrastructure to grow the towns and to accommodate the trades and merchants, the municipalities and the visitors travelling in between. It was also a legal construct administered by Holy Order. The merchants, towns and religious orders were fused within this symbolic legal structure.

The first stone bridge in 1172, the Judith bridge, was a link between the Knights of the Cross of the Red Star, the only Bohemian-founded religious order, and the Church in the Chain of the Holy Order of Malta. (What remains of this Romanesque bridge can be seen on the shore, north of the Old Town gate, and in the reintegrated Judith tower, the shorter of two on the Lesser Town side.) After the collapse of Judith bridge in a flood in 1342, the construction of a new permanent link was built. With advanced technology to resist both floods and their impacts to its foundations, Peter Parler began construction on the Charles Bridge in the year 1357. Construction and repairs continued for many decades after. It is said to have been started at 5:31, July 9th, 1357, at the numerically significant moment 135797531 (itself a symbolic bridge).

The new tower gates on both sides are 9.5 metres wide and 35 metres tall. The bridge was built with blocks of local sandstone, and was braced with 16 pier buttresses, over a total length of 515 metres. While today only 30 percent of the stones on the Charles Bridge are original, and the tower facade details have been reconstructed, the heights and form of the structure are original. Carefully engineered, diamond-shaped piers in the river support vast spans, between 16 and 24 metres, at a great height. They are protected by wooden constructions, rising out from the water, which were used as ice breakers to protect the bridge and to minimise pressures to the foundations.

The bridge is integrated into the social space of the city, becoming an extended merchant street between the Old Town and the Lesser Town. The Church in the Chain of the Holy Order of Malta, on the western side, had a privileged place within the city and controlled this critical connection as the guardian of the bridge. Rising from the ground in Lesser Town, the bridge jumps over Kampa island to span the river with subtle irregularity, before landing in a gentle, steep slope on the eastern side through the Old Town gate into the Knights of the Cross square.

At the end of the 30 Years' War, in 1648, the invading Swedish army was stopped on the middle of the bridge. After the war, Baroque redevelopments reflected the change of power and the renewed authority of the Church in daily life. Between 1706 and 1714, most of the 30 symbolic sculpted figures were added along the bridge (only 1 is still original, the rest are replicas). The Old Town side, which was dominated for three centuries by the civic tower of Charles Bridge, was transformed into a small and sacred square with the construction of the Knights of the Cross Church and the Clementinum library. In 1723, the Bridge was illuminated with oil lanterns, and in 1785 a staircase was installed to connect with Kampa Island. In 1905, an electrified tram crossed the bridge later to be replaced by buses. The traffic however disrupted the bridge's stability and in 1965 it became a pedestrian only path.

With its mixed Gothic and Baroque forms, its combined elements of Crown and Church, and its arhythmic framing of views over the Prague Castle and the landscape along a medieval merchant street, Charles Bridge captures the spirit of the entire city. In the early 20th century, plans for a highway that would have passed beneath the bridge along the shores of Lesser Town and Kampa island, was fortunately not realised due to the outbreak of the war. Today, the bridge is a place to watch street musicians and artists, sunsets, fireworks, and wintertime snow. The bridge is at its most beautiful when one is lucky enough to find it empty of people.

老城广场的旧市政厅
Old Town Hall

地址：Staroměstské náměstí 1/3, 110 00 Praha 1, Staré Město
交通：Metro A **Stop** Staroměstská（5 分钟步行）
建造时间：始建于 1338 年

□ 哥特/ GOTHIC
■ 文艺复兴/ RENAISSANCE
■ 巴洛克/ BAROQUE
■ 新哥特/ NEO-GOTHIC
■ 共产主义/ COMMUNISM

老城广场是布拉格最重要的城市公共空间，从 13 世纪初起便成为老城的贸易中心。广场周边围合着不同时代和尺度的建筑，随着地貌变迁，原有罗马风时期建筑的首层现在都变成了地下层。哥特式拱廊与原先独立的建筑一起连接成更大的建筑及建筑群，并覆以文艺复兴和巴洛克的立面。哥特式、文艺复兴、巴洛克式、洛可可、19 世纪晚期的新古典主义，以及 20 世纪初的新艺术运动等不同时期风格的建筑都环绕着广场。泰恩教堂独特的双子哥特式尖顶成为老城广场醒目的标志。广场中心还竖立着由拉基斯拉夫·沙劳恩创作的捷克宗教改革先驱扬·胡斯的雕塑，雕塑于 1915 年 7 月 6 日（扬·胡斯逝世 500 周年纪念日）揭幕。如今布拉格老城广场是欧洲最大、最美的"城市客厅"之一。

作为老城广场的地标，旧市政厅始建于 1338 年，容纳了古往今来的各种历史痕迹。从布拉格城堡的视角来看，旧市政厅作为一座市民之塔，标志着市场和城镇地位的不断攀升。这是当时最为先进和昂贵的工程，它用近 70 米的高度宣誓它的重要性，在此之前，最重要的城市景观由教堂和宫殿来承载。建筑垂直四角上用大石块砌筑加固墙体，防止扭曲和开裂。塔的东部是一座建于 1381 年的圣母玛利亚礼拜堂，也是捷克最珍贵的古迹之一，设计者为彼得·帕勒（圣维特大教堂的建筑师）。塔的南部则是著名的天文钟。

到 15 世纪后期，旧市政厅的扩建工程逐渐开始，因此向西面并入了更多的房屋和花园。这些不断积累变化的细节仍然可以在建筑南立面上清晰可见。在以后的 300 年，再建主要集中在建筑内部。塔楼顶部的标志性景观外廊于 1787 年增建，以庆祝四个城区（老城、小城、新城和城堡）统一成为一个城市，站在外廊上可以俯瞰美丽的城市全景。塔楼石板屋顶顶角的小金属尖顶突出了建筑的垂直性，尖顶上方装饰着用于反射阳光的黄铜球。为了便于攀爬，市政厅塔楼内经过现代化改造，设有完备的轮椅回廊道及电梯。

天文钟是市政厅辨识度第二高的建筑元素，欧洲最具人气的旅游景点之一。作为塔楼南面的附属物，从 1410 年起，

时钟就被装饰到塔上。今天看到的无比精致的天文钟安装于 1490 年。天文钟由星盘、日历盘和十二圣徒组成。日历盘上面的 12 张季节图片反映了波希米亚的乡村生活。整点时分，天文钟报时。铃声敲响，沙漏倒转，12 名圣徒经过天文钟上部的窗户，向游客点头致意。

在建筑内外观细节中，可以看到层叠的历史：哥特式、文艺复兴和巴洛克式以及近现代的修缮。建筑外部，哥特式和文艺复兴时期的石制窗框从多彩的巴洛克式和新古典主义风格的泥灰墙壁中探出。建筑内部，各层房间都呈现出其所属时代的建筑风格，建于 1470 年哥特式议会厅的彩绘墙与 20 世纪 50 年代重建的切割玻璃块砌筑的北墙连通，并有楼梯通向室内由混凝土浇筑的塔楼。

1830—1845 年，市政厅北部扩建为新哥特式风格。这部分在 1945 年德国纳粹军队逃离城市时遭到严重破坏，现在已经被拆除并成为一个小公园。可以从塔楼的东北角看到，作为一个建筑片段，这座建筑的北翼遗迹仍然保留着。这个遗存是布拉格唯一存在的独立建筑片段，与其他已经重新整合到新建筑物中的碎片不同，这个片段保存着一段"缺失"的记忆。

如今，老城广场除了 2020 年重建了在 1918 年被推倒的巴洛克时期的玛利亚圣母柱外，没有兴建任何现代建筑。关于如何填补广场被炸毁遗留的空地已经进行了多轮建筑方案竞赛。在过去的一个世纪里，几乎每一代人都曾尝试过，但迄今仍然没有产生什么结果，或许就这样空着会更好也说不定。

… at the centre of the oldest trading hub of the city, at the centre of the Old Town Square.

By the early 13th century, at a raised elevation just high enough above the periodic floods of the Vltava, this location was the ideal place for the centre of the trading town. Within the ancient natural topography and embraced by the bend in the river, it had been a meeting point of trade for centuries. Following the contours and the pathways of the land, the edges of the Old Town Square in fact are not "square"; they bend and curve as does the ground itself.

The primary civic landmark of the Old Town, the Old Town Hall, stands at the centre of the square. Wealthy merchant houses maintained the most privileged positions around this central point. The Gothic Church of our Lady Before Týn, on the other side of the square, constructed in 1380, is taller than the Town Hall but it is set back from the marketplace behind merchant houses. The church was to remain a supporting rather than a defining element of the Old Town Square.

The market developed through the rebuilding of its earliest settlements into an organic architecture of stone. Most of the buildings today have underground levels from the Romanesque period which had once been on ground level. The market was once composed of houses of a different scale than what is seen today. Gothic arcades, each originally a single house, over time combined into larger groups and larger buildings to be covered with Renaissance and Baroque facades. Structures from Gothic, Renaissance, Baroque, Rococo, late 19th century Neo-Classicism and Art Nouveau styles are found on the square.

Having been constructed, in time, along with the square itself, the Old Town Hall contains all the architectural styles which developed since its construction began in 1338 when King John of Luxembourg, father of Charles IV, granted the town his approval. Its urban footprint within the square, as well as its architectural styles, adapted to meet the needs of the growing town. From the point of view of the Prague Castle, the Old Town Hall — as a civic tower — marked the rising position of the market and its town. It asserted its significance at a height of nearly 70 metres, in a landscape which had until then been primarily dominated by Crown and Church. The large stone blocks which define the vertical edges of the tower reinforce the masonry walls against twisting and cracking — an advanced and expensive engineering work at the time.

The construction of the Old Town Hall and tower began with the purchase of a two-storey house and garden, from Welflin, a merchant. The gradual accumulation of buildings and expansions of the Old Town Hall began in the late 15th century, with the purchase of more houses and gardens to the west. The rhythm and articulation of these accumulations can still be seen on the subdivisions of the south facade. The Town Hall eventually integrated 5 properties. Reconstruction over the following 300 years was predominantly to the interior. The iconic viewing gallery, at the top of the tower, was added in 1787 to celebrate the unification of four towns (Old Town, Lesser Town, New Town and Hradčany) into one municipality. Small metal spires at the top corners of its slate shingle roof accent the structure's verticality, and are topped with brass spheres to reflect the sun.

The Astronomical Clock is placed on the southern side of the tower, with a small chapel from 1381 around the corner. The first clock was integrated into the Town Hall as early as 1410. The astronomical clock seen today was installed in 1490 in a highly refined Gothic style, equal in quality to that of the Prague Castle. It measures the sun, stars, planets and moon, and symbolises the governing universe. It demonstrates a clear distinction between heavens and earth, good and evil, and man and woman. The moving figures speak to the people every hour. In 2017, a renovation began to return the clock to its original construction, replacing metal gears with those made of wood.

Within the Old Town Hall complex and tower are various rooms, which together present the architecture of different times. In the architectural interior and facade details, an overlapping of histories can be seen (Gothic, Renaissance and Baroque through late modern renovations). Outside, Gothic and Renaissance stone window frames project from the colourful Baroque and Neo-Classical stuccoed walls. Inside, the painted walls of the 1470 Gothic hall of the Council Chamber is linked into the 1950s reconstruction of the northern wall, made of cut glass blocks, which lead one upstairs towards the cast-concrete interior of the tower, equipped with an elevator.

The Town Hall's northern expansion, in 1830–1845, was in a Neo-Gothic style. Severely damaged in 1945 by the German army as they were fleeing the city, it was demolished and is now a small park. The remains of this northern wing are still preserved — as a rough cut fragment — seen at the tower's Northeast corner. This broken edge is one of the only fragments in

Prague which stands by itself. Unlike other fragments which have been re-integrated into newer buildings, this one signifies absence and retains the memory of "the missing something".

Most recently, the Baroque-era Marianský column (demolished in 1918) was rebuilt in 2020, but there is no other contemporary building on the square. The ongoing debate about completing the northern wing of the Town Hall has gone through multiple competitions. Nearly every generation over the past century has tried but failed to produce a result — perhaps for the better.

文艺复兴时期

?—1526—?

这个时期也许起源于15世纪早期布鲁乃列斯基的透视结构中，或许起源于15世纪末期探险家们的制图模型之中，或源自1336年彼特拉克登旺图山的信件里，或在1420年7月14日胡斯信徒在维特卡山顶取得胜利标志着个人主义和宗教改革兴起时，抑或在1458年圣杯派贵族波杰布拉德的伊日成为第一位"胡斯信徒"国王时。但很明确到了1492年，商人阶层在帝国范围内广泛发展，并通过全新的建筑形式对城市产生影响。来到布拉格的意大利人影响了布拉格的文艺复兴，这种影响在哈布斯堡家族的费迪南德一世取得波希米亚王权之后更为显现。1537年至1576年间波希米亚的文艺复兴达到了最高峰，但远晚于欧洲的同一时期。

随着时间的推移，资本和权力的集聚以一种非常特殊的方式重新被赋予到城市景观之中。经济和政治的自主权相继赋予了人民，特别是与市场相关的商人阶层，这一举措推动他们建立一个旨在关注自身历史进程以及可控未来发展的"独立"的中产阶级社会。他们由犹太人、斯拉夫人、拉丁人以及日耳曼人等组成，并分化产生了罗马天主教、胡斯派、路德派等信仰。这时阿拉伯数字系统被引入，人们采用笛卡尔坐标体系以及透视来认识空间。

在古代探索发现以及征服新疆域活动的驱使下，等级制度的绝对逻辑与自我的个人主义逻辑相结合，形成了一种新的世界观：理性以及自我意识作为一种社会价值符号得到体现。建筑开始有意识地表现出主人的身份以及信息。建筑图像从象征主义和圣像转换成一种使用理性以及均衡的数学形式（如黄金分割比例）进行表达的新的视觉抽象。

对于罗马风的参考，不再只是通过单跨拱的建造来体现，而是通过对古代图像、元素和符号更加复杂的重建而实现。建造者的想法、地位以及标志被结合到了建筑立面中。现被视为"非理性"的过去的建筑外观按照新的几何法则简化而进行重建。与随意的哥特式形制不同，文艺复兴时期的建筑具有很强的逻辑性，即在于同化和吸收不同的要素，并将较小的单位要素合并成一个更大的整体。

完美的几何图形如正方形和圆形，被用来制造这些有着完美秩序的图景。这些"真实"和"纯粹"的语言也将希腊-罗马的词汇与数学和测量学方面的

文艺复兴时期的布拉格
The Renaissance City

新发展结合，去展现一种有内在固有知识、全新且充满想象的形式。为实现这种理想的平衡，从绘画以及建筑中可以看出，文艺复兴时期的平面图画代表了一种从一个固定的视点窥视"真实世界"的画面。凭借这种新的科学表现技术，人类意识到了他们面对的平面画布中三维空间的要义。通过透视的表现，二维的人物形象栩栩如生。

胡斯信徒反抗已建立的天主教秩序，引发了社会改革和波希米亚人对国家政治生活的更多参与。被这种新兴的等级制度和个人主义所强化，一个独立于帝国、教会和中产阶级的强大贵族阶级，成为这个新兴的商人阶级社会中的调

五彩拉毛粉饰外立面
Sgraffito Facades

彩绘木顶
Painted Wooden Ceilings

解人。在查理四世之后的两个世纪里，布拉格和维也纳之间的对于哈布斯堡王权的转移，使得地方当局能够重新集中他们的财富和权力，而叛逆的精神一直暗涌着。

人文主义与理性主义的出现同样也将这样的具有象征意义的宫殿建筑带入城市。文艺复兴时期作品的特征是通过将"古代"的符号转变为新的"理性"形式来进行应用和表现自己在历史和时空中的形象。这种具有代表性的想象将几何和象征图形与政治情形融合，并在整个社会中采用，在那个时期的房屋、宫殿、教堂甚至室内都能够看到。艺术家们更多地把眼光投向人的本身和日常生活，描述世俗生活的乐趣，充分体现出自然界和人体的魅力。布拉格文艺复兴的建筑是轻快和透气的，常常有敞开式的拱形长廊和修长的廊柱，并在外墙上引入意大利的拉毛装饰技艺。

下面的例子可以看作是文艺复兴时期（早期和晚期）理想建筑的模型。

THE RENAISSANCE ?–1526–?

... perhaps it began in the early 15th century with Brunelleschi's perspectival constructions, or in the cartographic modelings of late 15th century explorers. Or perhaps in 1336 with Petrarch's letter from Mount Ventoux. Perhaps the rise of individualism and reform was marked on July 14, 1420 with Žižka's Hussite victory on Vítkov Hill, or when Jiří of Poděbrady became the first Hussite king in 1458. Certainly by 1492, a decentralising society of the merchant class, upon a vast international scale of empire, made impact on the city through an entirely new form of representation in architecture. The arrival of the Italian influenced Renaissance, through the House of Habsburg, with its seat in Vienna, as its primary importer, had its peak between 1537 and 1576 in Bohemia, quite late by European standards.

With time, an accumulation of capital and power was re-distributed into the landscape in a very particular way. The continued granting of economic and political autonomy for the people, specifically the merchant class associated with the markets, established a new "free-standing" middle class society aware of its own historical progress and calculable future. Roman Catholic, Hussite, Lutheran, Jewish, Slavic, Latin, and Germanic: the people of Bohemia became more particular. A new numerological system was adopted (1, 2, 3...) which — unlike the Roman numerals — was capable of reaching the infinite. The Cartesian system and the perspectival governed space.

Motivated by ancient discoveries and conquests of new territories, the absolutist logic of the hierarchical combined with the individualistic logic of a self, to form a new conception of the world, in which a rationality and a self-consciousness were exhibited as a socially valued sign of status. Architecture began consciously to represent the identity and the message of its owner. Imagery moved away from symbolism and icons, to a new visual abstraction, which used rational and mathematically balanced forms (such as the golden ratio).

Roman references were no longer built through the single-spanning arch alone, but via a more complex reconstruction of imagery, elements and symbols of ancient times. Ideas, status and signs were combined in the making of the facade. Facades from past times, now viewed as "irrational", were reconstructed by simplifying and applying an order according to new—

涡卷
Volutes

重新应用古代的符号
Reapplied Ancient Symbols

体现社会价值的图像
Images of Social Values

几何的精度
Geometrical Precision

and geometrically definitive—laws. The "medieval" formula of symbolic depictions was transformed here into a meta-narrative of true form in itself.

From the undifferentiated masses of the Romanesque, the low and high reliefs of Gothic figures emerged and evolved into free-standing and ultimately self-defining forms. The figures of the Renaissance facade become signs and indicators of higher meaning. Unlike the haphazard Gothic, the demonstrable logic of the Renaissance image was in its absorption of dissimilarities, merging smaller units into a larger cohesive whole. The Renaissance wall itself was without significance — just bricks. It was to become a coherent symbol of order — a representative picture of sophisticated meaning.

Ideal figures, such as squares and circles, were used to make these pictures of perfect order. The "pure" and "true" language also combined Greco-Roman vocabulary with new developments in mathematics and measurements to demonstrate a newly imagined form of inherent knowledge. To achieve this enlightened balance, seen in painting and architecture, the Renaissance image represented a picture of the "true world" from a single and fixed point of view. With this new scientific technology of representation, humans became aware of the significance of the three dimensionality of the flat canvas they were facing.

With perspectival representation, came the full independence of the figure from the wall. Simultaneously with its physical autonomy, came its individual consciousness. The static position, and the mental and cognitive space of perception established a new system of order which encouraged all people to take part in it. It was based upon new rules of social representation and the collective image of an open field of hierarchical flexibility. It became the symbolic form of the free market space.

The emergence of Humanism and Rationality also brought the symbolic palace into the city. Seats of noble power positioned their image into the lands of their choosing and in the manner of their choice. Within this new world, a clear hierarchy of scale was established between people and the places they occupied. This had an impact on the daily business of townspeople as well. Each was competing to emulate most successfully the symbolic language of these highest up (on the sacred hill, so to speak). The figure (symbolic) and ground (topological) were inseparable in this new paradigm, as were subject and spectator. Renaissance representation assumed the conscious awareness of the position of the freestanding figure on its own ground — always, however, within the larger pre-determined field.

The Renaissance project of representation consisted in applying and representing an image of oneself within the historical and spatial world, by adapting "ancient" signs into new "rational" (and therefore demonstrably higher) forms. This representative imagination involved the merging of geometrical and symbolic figures with political status and was adopted throughout the society, as can be seen in houses, palaces, temples and even rooms.

The Hussite rebellions against the established Catholic order initiated both a reform of society and a new effort to manage the increasingly individualistic people. A strengthened aristocracy, independent from the Empire, the Church and the middle classes, was reinforced by this emerging order of hierarchy and individualism, and developed its new role as mediator in the society of the rising merchant class. During the two centuries after Charles IV, the shifting of the Crown as the Hapsburg heart, between Prague and Vienna, allowed local authorities to re-concentrate their wealth and power, while the rebellious spirit persisted beneath the surface.

The following examples can be considered as private and public representations of the (early and late) ideals of the Renaissance.

6. 安娜王后的夏宫
Queen Anne's
Summer Palace

7. 施瓦岑贝格宫
Schwarzenberg
Palace

安娜王后的夏宫
Queen Anne's Summer Palace

地址：Mariánské hradby 52/1, 118 00 Praha 1, Hradčany
交通：Tram 22, 23 **Stop** Královský letohrádek（1 分钟步行）
　　　Tram 22, 23 **Stop** Pražský hrad（7 分钟步行）
建造时间：1538—1563 年
建筑师：保罗·德拉·斯特拉（Paolo della Stella），乔瓦尼·迪斯帕齐奥（Giovanni di Spazio）等

☐ 文艺复兴/ RENAISSANCE
■ 巴洛克/ BAROQUE
■ 新哥特/ NEO-GOTHIC

布拉格城堡往北，穿过鹿苑城壕就到达皇家花园。安娜王后的夏宫建在这个曾经是葡萄园的皇家花园的东端。它优雅地矗立在一个平台上，俯瞰城市，并遥望着布拉格城堡，优美的景观尽收眼底。

15世纪末，文艺复兴风格开始渗透到波希米亚，先时的哥特风格开始慢慢与文艺复兴元素融合。随着大批意大利建筑师的涌入，文艺复兴风格在波希米亚被广泛接受，这在当时空间宽阔的城堡、带拱廊的优雅庭院和排列成几何图形的花园中得以体现。建筑设计的重点转向舒适，娱乐休闲建筑也开始出现。安娜王后的夏宫就是早期意大利风格，由费迪南一世在1538年为他的妻子安娜·贾吉隆建造，被誉为意大利境外最纯粹的文艺复兴风格建筑。夏宫由意大利建筑师及石匠保罗·德拉·斯特拉设计，最终由建筑师乔瓦尼·迪斯帕齐奥完成建造。

作为皇家休闲和避暑地，夏宫并不以实用性为首要目的，主要体现特权和社会地位，因此它以一种独立的姿态矗立在城市空间中，以贵族阶层希望的完美形式出现。建筑风格简约，线条优雅，体量纯净。建筑在各个方向上均衡，并在所有视角达到平衡。各个面的空间平衡也减轻了建筑的重量感并增强了屋顶的轻盈感。上层窗户的长方形和圆弧形图案虚实交替，均匀变幻的节奏令人赏心悦目，以一种完美的形式展现着轻盈和力量。

建筑一圈弧形的铜屋顶像篷布一样紧绷在宫殿高大而纤细的石柱围廊上。精巧的柱廊作为一种轻巧、纤细的外部围合，似乎失去了它的结构功能，纯粹成为一种象征，就像一个降落伞一样，外观柔软优雅，展示出一个有序而平衡的空间。宫殿内墙被描绘神话、狩猎和战争的浮雕装饰得富丽堂皇，顶棚用波希米亚王国的象征符号装饰。一楼有舞厅、画廊和天文台。

夏宫的前面存留下了建于1564—1568年文艺复兴风格的歌唱喷泉，青铜锻造，是铸造大师托马斯·亚洛斯（也是圣维特教堂最大的钟——西吉斯蒙德的制作者）的魅力之作。它是布拉格现存最古老的喷泉。喷泉被安置在石头水池中，

落在下碗边缘的水滴仿佛在演奏音乐。喷泉的中间描绘了一个希腊牧神——潘（森林和溪流的保护者），喷泉顶部站着一个小吹笛者。

几个世纪以来，皇家花园已经发生了许多变化。在三十年战争中被瑞典军队洗劫一空，之后被改造为种植异国情调植物的巴洛克花园，到了19世纪又被改造为英式的新古典主义园林。在1846年受新古典主义影响重建之后，夏宫的内部细节遭到损坏并被重新设计。在20世纪50年代，建筑的基础、屋面及廊柱得到了修缮。尽管夏宫的室内及花园已被改变，但建筑的外观还是保持了最初的面貌，从那时起就被用作展览空间。

... on its own plinth, it stands upon a raised ground atop a hill within the height map of the city, to occupy a symbolic space of elevated privilege.

A royal, recreational retreat was placed between the towns below and the Prague Castle above. In a style imported from Italy, it was meant to represent the spatial and social position of privilege. With references to the Roman temple and the acropolis, from a distance, it symbolised hierarchy itself.

It is often called the most beautiful Renaissance palace north of the Alps. Begun in 1538 by Ferdinand I for his wife Anna Jagiellon, it imported an earlier Italian style. According with the spatial theories of the time, it was designed by Giovanni di Spazio and Paolo della Stella (whose names mean "Space" and "of the Star"). As an exemplary prototype of the Renaissance Palace, it is a recreational place for Royal activities, looking upon the city and looking up to the

Castle. Due to the weather in Prague, it was to be used during the warmer months as a retreat, having the same qualities yet being more accessible than Italy. Built mostly as a symbol and a model of perfect luxury, it did not need to be useful, but be a pure representation. Queen Anne did not live to see its completion as she died in 1547 after giving birth to her 15th child.

North of the Castle grounds, a new wooden roofed bridge was built in 1535, to cross the Deer Moat into the new Royal Gardens. Queen Anne's Summer Palace was placed at the eastern end of these expanded gardens at a place which was once a vineyard. Within these walled grounds and past the Orangery, inside its own pleasure garden, the Summer Palace stands on its own platform to heighten its view over the landscape 50 metres below.

On its plinth of stone, this figure in the round is capped with a copper roof. Looking backwards to the art of Quattrocentro Italy, it made no attempt and showed no interest to assimilate into the towns or its present. The corner columns recall the Florentine frescos of the early 15th century, such as Masaccio's Tribute Money in 1425. With equality on all its sides and in perfect balance from all views, it demonstrates the isometric truth of form itself. This spatially fluid wrapping around itself lightens the overall mass and enhances the perception of the weightlessness of the enormous roof. Almost like a parachute, its appearance is soft and graceful.

A space of power, subtlety, peace and nature, it is a form of perfection for the privileged user only. The architectural minimalism, a purity of line and volume, describes a clearly ordered universe, measured and in balance. It demonstrates a hierarchical and symbolic reasoning of a perfect form of lightness and strength. The alternating pattern of squares and circles on the upper level windows, pleases the eye with its even, oscillating rhythm.

As if hanging in tension, on tall and thin Ionic columns, the palace seems to be pinned into the earth. As a lightweight, finely detailed and thin wrapper, the mathematically precise colonnade appears to lose its structural function, becoming pure image. The structure is an imaginary wall — with its inner volume set back behind the lacy weaves of circular arcs. The Palace was richly decorated with wall reliefs depicting mythology, hunting and wars, and the roof was detailed with symbols of the Bohemian kingdom. A dance hall, gallery and astronomical observatory were on the first floor.

Over the centuries, the garden has seen many changes. The game fields, maze and aviary of the time have since been replaced. Among the remaining features of the garden are the Renaissance fountain from 1568. The Palace was ransacked by the Swedish army in 1648 and after the 30 Years' War, the gardens were renovated in a Baroque style with exotic plants, and again in the 19th century as Neo-Classical gardens in the English manner. As it appears today, the trees are taller and partially limit the view, which would not have been the case originally. After Neo-Classical reconstructions in 1846, the interior details of the Summer Palace were damaged and redesigned. In the 1950s, there was a complete reconstruction of its foundations, columns and roof. Despite these changes to the gardens and the interior, its original exterior appearance remains preserved.

施瓦岑贝格宫
Schwarzenberg Palace

地址：Hradčanské náměstí 2, 118 00 Praha 1, Hradčany
交通：Tram 5, 7, 11, 12, 15, 20, 22, 23, 41, 97 **Stop**
　　　Malostranské náměstí（12分钟步行），Pohořelec（6分钟步行）
建造时间：1545—1567 年
建筑师：阿戈斯蒂诺·加利（Agostino Galli）

□ 文艺复兴/ RENAISSANCE

施瓦岑贝格宫坐落在布拉格城堡入口处的赫拉德坎尼广场上,是一座重要的文艺复兴时期建筑。1541年的一场大火摧毁了广场上的大部分建筑,1545年,洛布科维茨家族委托意大利建筑师阿戈斯蒂诺·加利在此建造了这座宏伟壮观的建筑,成为赫拉德坎尼山上规模仅次于布拉格城堡的宫殿。

施瓦岑贝格宫展现出一种文艺复兴晚期的矫饰风格,这也是巴洛克式风格的先导:通过自身可塑性来拉动空间本身。建筑拥有巨大的檐口,精致的砖砌工艺使拱顶仿佛能够弯折,从而将顶部挑出更远。这些上层飞檐使得站在底层的人抬头向上看时感受到一个更为庞大的建筑体量。通过将基地映射到其自身的结构框架中的方式,更进一步地将建筑融入城市以及社会空间中。

五彩拉毛粉饰是施瓦岑贝格宫最重要的波希米亚文艺复兴风格装饰特征。这是一种从意大利引进的将图形蚀刻入墙壁形成复杂空间形态的技术。这种透视感的建筑表现创造了一种具有深度以及活力的虚幻效果。宫殿表面被精美绝伦的五彩拉毛装饰包裹,乍一看,立面上的每一个砖块好似都能被人在心理上感受到庞大的质量感,从而让人立即感受到这个建筑庞大的规模体量。

宫殿因其庞大的体量支配着周边的环境,建筑立面在很远的地方都可以吸引人们的注意。由于宫殿建在峭壁上,因此从两侧显示出不同的层数。在宫殿高大的山墙上,蜗形和拱形用于弯曲和软化边缘,并遮盖屋顶。檐口和壁柱的连接处由纯手工精制,它是从意大利引进的文艺复兴的特征,但同时也被染上了波希米亚的风格。施瓦岑贝格宫可能是波希米亚与意大利风格融合最为精巧复杂的一个例子。

在宫殿南部的烟囱上可以看到一个金色的日晷。在日晷的两侧分别是一个公鸡和猫头鹰的形象。作为清晨与智慧的象征,它们与宫殿的象征意义相呼应——在太阳和地平线的交会点呼唤着新的一天的来临。这座宫殿在岁月变迁中始终保持着与城堡并驾齐驱的威严。

今天,施瓦岑贝格宫是捷克国家美术馆的一部分。宫殿内许多房间现在仍然装饰着当初文艺复兴时期的挂毯,古希腊主题的彩绘天花也保存完好。在这里,将有机会欣赏波希米亚文艺复兴和巴洛克艺术的丰富收藏。

另一个同期的文艺复兴建筑:格拉诺夫斯基宫,位于老城泰恩庭院,这里原是贸易中心和外国商人的海关,也非常值得一看。

... on the upper threshold between the merchant city and the gates of the Prague Castle.

Another example of noble power, the Schwarzenberg Palace created a different dynamic from Queen Anne's Summer Palace. In its relation to the Castle, it was the pinnacle of society, overlooking the entire landscape from its fortified walls. The Schwarzenberg Palace is a primary building element of the campus which emerged on Hradčany square over the 16th to 18th centuries. It is one of the composite of structures representing the leading authorities of power — the Crown, the aristocracy, the Church, the market and the leading merchant houses.

After the fire of 1541, which destroyed most of the buildings on Hradčany square, there was an opportunity to rebuild. Rather than rebuild the typical houses of a market, in such proximity to the Castle, it was possible to establish a more significant representation of social and topological standing on this site. The Lobkowicz family began construction of a palace of exceptional grandeur in 1545, by Italian architect Agostino Galli. The site was a T-shape plot of two lots with others added in 1800.

The T-shape building faces both the Castle and the town, and is intended to be seen by both, as well as from the street. Its standing is immense. Due to the cliff on the one side, it is both a 3 story and 8 story building at the same time. On the side of Hradčany square, it is a gated palace with a forecourt framed by a wall. Its inner courtyard is partially visible through the metal gateway as an extension of the market, albeit a private one. On the steep terrain of the south side, the Palace is built on massive fortified walls. The decorated facade of its upper levels is clearly differentiated from this lower, unadorned structural mass.

Sgraffito, the technique of etching imagery into walls, was imported from Italy as a sophisticated science of spatial and graphical form. In a play with perspectival perception, it creates the illusory effects of depth and vibrancy. The extraordinary sgraffito facade, covering an area of 7000 square metres, wraps the skin of the massive walls in a delicate eruption of detail. In an instant, each unit can be mentally quantified as a total mass, immediately creating the perception of vast scale.

Dominating its surroundings, the building's facade demands attention from a great distance. On the tall brick gables, volutes (the fusion of square and circle used to hide a triangle) and arches are used to curve and soften the edges and to hide the roof (an inelegant technical necessity to be masked). The articulations of the cornices and pilasters, refined with the control of a hand, were the signs of a Renaissance language imported from Italy and coloured with a Bohemian touch. They combine the forms of a Gothic lancet style with the techniques and themes of a late Italian Renaissance. Schwarzenberg Palace is perhaps the most sophisticated example of this fusion of Bohemian influence and heritage with Italian style.

The exterior massing has a large cornice which claims volume beyond its property lines. It displays a late Renaissance, mannerist quality of grand splendour. With its exaggerated yet reasoned forms, it can be considered Mannerist — the pre-cursor to a Baroque plasticity — pulling on space itself. Refined brickwork allows the vaults to curve with crisp edges, far into the space overhead. These upper cornices acknowledge the scale of a figure standing on the ground below looking up, framed by this higher ground. This reflection of the ground into the frame of the building further integrated it into its urban and social space.

On the southern chimney of the Palace can be seen a golden sundial. On either side of the sundial, there are a rooster and an owl. As the symbols of the morning and of wisdom, they echoed the symbolism of the palace — at the intersection of the sun and the highest horizon of the city, calling forth the new day. The Palace has changed hands for centuries, yet has always retained its prestige alongside the Castle.

Today, the Schwarzenberg Palace is part of the National Gallery. It features many exhibitions that include the interiors of the Palace themselves. Many rooms are still adorned with original Renaissance tapestries and painted ceilings of ancient Greek themes.

Another example of Renaissance architecture in Prague built at the same time, is the Granovsky Palace within Ungelt (or Tyn Courtyard), a fortified merchants' yard located in the Old Town.

巴洛克时期

?—1618—?

……这个时期也许起源于1412年扬·胡斯到达山羊堡开始传道，或许起源于1515年马丁·路德在德国维滕贝格的宗教改革，或许可能产生于伽利略在1610年发布他最初利用望远镜观测天体的结果时，也或许产生于鲁道夫二世作为波希米亚国王和神圣罗马帝国皇帝在1583年将哈布斯堡帝国首都定在布拉格时。当然，也可以说起源于1618年在布拉格引发的三十年战争开始时，这时社会内部的矛盾激起。

新教改革是社会历史意识以及个人主义兴起的开端。天主教及皇权等受到挑战，随之绝对平衡被打破，一种新的动态显现出来。这时，波希米亚是神圣罗马帝国最大且最为混合多元的地区。随着代表社会各阶层的地方和民间力量崛起，形成对过去既定社会秩序的一种挑战，各方面的影响因素使得社会的多样性和复杂性骤然提升。新建立的政权最终被制度化为一个全新的城市政治结构。这种具有多个中心的模糊性被吸收到建筑的雕塑性动态空间中。

巴洛克风格特异且迷乱的特性体现了当时社会的内部矛盾及紧张局势。改变和演化的可能性被表现在建筑形式中，建筑的立面不再是静态和平坦的。随着对历史开始全新认知，文艺复兴时期写实的形态描绘演变成为一种看似不和谐的光影表现。

布拉格在三十年战争中受到了很大影响。在1621年波希米亚人战败后，布拉格不再是皇权的所在地，再次沦为哈布斯堡王朝的一个省份。清晰的巴洛克风格在此期间开始渗透到波希米亚，最终取代文艺复兴风格。来自意大利天主教的巴洛克风格得到了波希米亚富有的天主教教会和贵族阶层强烈支持，天主教教会于1627年后成为唯一的合法教会，早期的巴洛克建筑师也都主要是意大利人。

在巴洛克的开始阶段，一种夸张的手法主义在试探着自身的极限。在整个巴洛克时期，这种投机和反传统的建筑风格在不断发展。巴洛克风格从最初的破坏性、多变性和剧烈性的象征表达开始，随着其活力的逐渐减弱和扁平化，又重新趋于稳定。

巴洛克时期的布拉格
The Baroque City

　　巴洛克风格的可塑性伴随着倾斜视线的出现、运动的物体、观念的叠加、对立的感知和体验，以及认知失调，这些都是对文艺复兴的平衡秩序理论的反驳。新的建筑表征采用了破碎和变形来加强对新的世界秩序的认知。三十年战争后，欧洲大地遭受了极大地破坏，以前被压制的力量又重新出现，这些新兴量在新出现的几何张力中得到回应：椭圆形、双列柱、破碎的山墙、平滑的运动以及多层壁柱。这些高度雕塑化的表达方式使墙的确定秩序和体量的绝对特征都受到质疑。

椭圆的形式
Elliptical Forms

透视错觉
Perspectival Illusion

社会秩序在相互冲突的势力日益激增的对峙下，产生巨大破裂，后逐渐平息并稳步发展。单一的秩序此时不再能够遏制社会的动荡，需要寻求新的制约形式。到 18 世纪晚期，在布拉格的巴洛克风格结束之时，也就是在洛可可风格晚期，建筑立面由最初的雕塑和立体感重回平坦的形式。巴洛克早期破旧立新的力量最终被吸收到一个更加稳定且确立的建筑表达方式之中，并持续了一段时间。

巴洛克风格是在布拉格及波希米亚得到广泛运用和流行的、最有特色的风格。"布拉格的巴洛克"风格也闻名于世。以下的例子可以被看作巴洛克风格早期到晚期发展过程，从其高度动荡性和雕塑感的形式逐渐趋于平静和扁平化，再达到一种动态稳定的状态。

双柱
Doubled Columns

高浮雕人像
Highly Sculptural Figures

分层壁柱
Layered Pilasters

断裂山花
Broken Pediments

THE BAROQUE   ?–1618–?

... perhaps it began with Jan Hus and the Hussites in 1412, or with Martin Luther in Wittenberg in 1515. Or perhaps when Galileo published his scientific critiques of the established laws, in 1610. Or perhaps when Emperor and King Rudolf II's ideologically ambivalent reign moved the Hapsburg court back to Prague in 1583. Certainly by the beginning of Europe's 30 Years' War, in 1618, the internal contradictions surging within society accumulated a massive social rupture within all the established institutions.

Contradictions of the period emerged in many forms. The Hussite (or Protestant) Reformation was the beginning of a time of socio-historical consciousness and rising individualism. The higher laws of Crown and Church were challenged. With the putting of the establishment out of absolute balance, a new dynamics was emerging. At this time, Bohemia was the largest and most mixed region in the Holy Roman Empire.

Upward mobility began to pose a threat. Catholic churches were converted, lands were confiscated, and power hierarchies within the towns were reconstituted. The economic tensions increased between the social classes in an eventual acknowledgment of local power constituted by the people and their groups. This culminated in a displacement of the Church's dominance and its subsequent adaptation within the grounds of the market.

The rise of local and civic authorities, represented by a varied base of people, challenged the established orders of the past, with diversity and complexity coming from many directions. Recently established powers were eventually assimilated and institutionalised into a new political structure of the city. Absolute order was put into question by an emergent and destabilising force: doubt. The ambiguity of having multiple centres was being absorbed into the architecture of a sculptural space of movement.

The strange and disorienting forms of the Baroque make manifest these internal contradictions and tensions of the time. The possibility of change and transformations were now represented as their own architectural forms. The facade image is no longer static and flat. The late Baroque acknowledged the presence of a cognitive subject, of a thinking person moving in the world. With the beginnings of a new consciousness of the historical condition, the Renaissance portrayals of harmonic form morphed into an optics of dissonance.

8. 华伦斯坦宫和花园
Wallenstein Palace and Gardens

9. 克拉姆 - 葛拉斯宫
Clam-Gallas Palace

10. 圣尼古拉斯教堂（小城）
St. Nicholas Church (Lesser Town)

11. 斯特拉霍夫修道院及其图书馆
Strahov Monastery and Its Library

During the period of Prague's Counter-Reformation (or Catholic Revival), after the controversial reign of Rudolf II, much of the Bohemian powers continued to oppose the Catholic Emperor. The 30 Years' War (a battle of ideas), a conflict which began in Prague and engaged most of Europe, — strongly impacted the city. Prague and Bohemia, no longer the seat of the Crown, were again, after their defeat in 1621, reduced to a secondary Hapsburg province.

Catholic churches, under the re-empowered Jesuit order, re-asserted their position in the defeated city. The newly re-established powers imposed forms of order into this moving space of the city: statues upon Charles Bridge, new dominants on Charles Square and Old Town Square, to name a few. The Church began to re-engage the fluid space of the market and the attention of a society of increasingly independent and competing individuals. While still facing eastwards towards Jerusalem, the churches now oriented their entrances to the streets and to the mobile society with its changing perceptions.

As the Baroque began, in its early Mannerist phase, displays of exaggerated forces were testing limits. Throughout the Baroque, there was a continuous development of speculative and iconoclastic architecture. The relaxing of its sculptural rhetoric followed from disruptive, volatile and violent symbolic representations, with a gradual diminishment and flattening of this dynamism and its reintegration into a stable establishment.

The plasticity of the Baroque, with the emergence of an oblique view, the moving subject, the overlaying of forms, the perception and experience of oppositions, and the moment of cognitive dissonance, became the antithesis of the balanced orders of the Renaissance. New architectural representations which used broken and twisted symbols were introduced to reinforce the new (dis)order of the world.

By the end of the 30 Years' War, European populations and landscapes were devastated. Destruction, disease and starvation were widespread and previously suppressed powers were emerging. These emerging forces were echoed within the emerging geometries of tension: ellipses, doubled columns, and broken pediments, moments of perceptual slippage, and multi-layered pilasters. Both the definitive order of the wall and the absolute character of the mass were put into question by this highly sculptural expression.

Massive rupture and then eventually calm developed through the increasingly dynamic play between the conflicting surges of powers. The pulls of the society could no longer be contained within a single order. New forms of restraint were sought. By the end of the Baroque in Prague in the late 18th century, in its late phase Rococo, we find the ultimate re-flattening of the sculptural facade. The iconoclastic power of the early Baroque was absorbed into a more stable and established rhetoric — which lasted for some time.

The following examples can be considered as a sequence through the developments of the early and late Baroque time, from its highly disruptive and sculptural form, through its gradual calming and flattening, towards an emphatic and reverberating stability.

华伦斯坦宫和花园
Wallenstein Palace and Gardens

地址：Letenská 123/4, 118 00 Praha 1, Malá Strana
交通：Metro A Stop Malostranská（3 分钟步行）
建造时间：1621—1630 年
建筑师：乔瓦尼·皮耶罗尼 (Giovanni Pieronni)

□ 巴洛克/ BAROQUE

华伦斯坦宫掩映在布拉格城堡的山脚下，其布局具有与城堡相似的制式，从城堡可以俯瞰该宫殿及花园。宫殿的主人是白山战役后波希米亚最具权势的贵族及将领，华伦斯坦家族的阿尔布雷希特·冯·华伦斯坦（1583—1634）。该宫是他聘请建筑师乔瓦尼·皮耶罗尼为其建造的一组新意大利风格的建筑。为了新建宫殿，原有基地上众多的住宅、作坊及客栈等老建筑等在1621—1627年间被拆除。建筑分阶段进行建造，并获得了一种混合但有凝聚力的形式。从城市街道来看，建筑群呈围合式，庭院仅向内部开放，反映出权贵和民众之间的分离。

华伦斯坦宫的三座主要建筑被围墙包裹着：两座主宫殿均位于基地的上方，围合式布置，并相互连接，面向城市空间；另一座位于东部的建筑是马厩。宫殿外墙呈现出一种不寻常的形式：长约750米、高8米，环绕着内部花园并将不同属性建筑组合成的一个更大整体。围墙内是一个娱乐天堂，有花园、骑马场、室外音乐厅、古神话雕像、温泉浴场、石窟，并饲养野生动物。同时，所有进入花园的入口都精心布置在角落，可沿着花园路径的对角线延伸空间的感知长度。在这片伸展的围墙里，即使是日落的投影也被放大了。在宫殿东侧的马场上，视线被特殊处理，以维护隐私，使得从布拉格城堡不能看到训练演习。这座建筑和它的主人，在如此精心地展示中，注定要从不同的角度观看。

萨拉·泰雷纳是亭子、露台和大厅结合的半露天剧场。它围合出三个等量的半室内空间。巴洛克风格的双柱（本身也是结构力量的象征），华丽而细腻的装饰形成了舞台的前景。有孔雀漫步的花园中轴线与剧场中心对齐，在不规整的场地中建立一种平衡的秩序。花园的南侧是一堵长170米、高8米的仿钟乳石墙，墙上有令人诧异的灰色纹饰，扭曲的人脸和野兽的形象出现在洞穴巨大粗糙的表面，进入其中是石窟和私人享用的温泉。

宫殿内的房间相互连接，每间房间都个性突出并装饰华丽，采用了大理石、木作、枝形吊灯、墙饰、皮革墙纸、镜子、水晶和彩绘天花，所有这些都突出了华伦斯坦的神秘力量。一个特别的房间是主宫殿内的大厅，通往这个空间的门和楼梯高大宽敞，可骑马通过，彩绘天花和镜子也进一步扩大了空间的视觉效果。另一个特别空间是华伦斯坦的私人小教堂，这是宫殿内一个隐秘的高耸空间，里面有绚丽的彩画和金饰浮雕，靠近顶部的一扇小窗户朝向布拉格城堡。

如今，这座宫殿是捷克共和国参议院所在地，同时是国家博物馆和教育部的一部分。花园向公众开放，可举办展览和音乐会。

... in the heart of Lesser Town, at the foot of the Prague Castle, a counterweight to the scale and symbolic power of the Crown.

Prague Castle and Wallenstein Palace are approximately the same size and shape. The layout of the Palace is like a fortress. Courtyards in the buildings of the main halls open inwards, guarding at all times against a possible breach. From the point of view of the city streets, there was a controlled separation between the Palace and the people of the city.

The new power established with the defeat of the Czech Estates at White Mountain in 1620 made an aggressive effort to express a completely new and dominating order. Wallenstein, an upward rising aristocrat and warrior (who was later killed for his ambitions) was at the centre of power struggles in Europe during this time and established his domain in this new Estate. The older buildings on the site, inns, artisanal and residential structures that had accumulated since the 13th century, were completely demolished between 1621 and 1627. During 1624–1630 the construction of the palace and garden began as an absolutely new imposition into the site of 26 former properties. It was realised in phases and acquired a mixed yet cohesive form.

Despite its position at the base of the city ground, the Wallenstein Palace in fact presumes the status of the Castle; the Castle, in its privileged perch, must look into the gardens and see the power and wealth that Wallenstein had accumulated. Wallenstein employed architect Giovanni Pieronni to import a new style of Italian architecture for his Palace and Garden.

Three main buildings were built within an enclosing wall: the western palace facing the space of the city, its extensions on the north, and the horse stable on the east. As a direct consequence of the combination of separate properties into a larger whole, the exterior walls delineate an unusual form, irrational and unexpected. A nearly 750-metre-long, 8-metre-high wall wraps the internal gardens.

Inside the walls was a recreational "paradise" with plants, riding grounds, an outdoor concert hall, ornate statues of ancient mythology, a thermal spa, a grotto, and wild animals. The architecture and the man, in a deliberate display, were meant to be seen and to be looked at from varying angles. At the same time, all points of entry into the gardens were controlled. They appear at corners, are approached obliquely, and extend the perceived length of the space along its diagonal. Even the sunset itself was exaggerated in this stretched enclosure. At the riding grounds on the eastern side of the palace, the mastery of perspectives was used inversely, to maintain privacy and block the training exercises from the Prague Castle's view.

The Sala Terrena, an open-air theatre, exhibited the morphology of a pavilion, a terrace and a hall combined into an outdoor setting. It displayed Baroque motifs of double columns (themselves a symbol of excessive structural power) to partially enclose the "interior space" of three equal bays. Highly ornate and detailed optical games of imagery formed a backdrop to the stage which opens to the garden where peacocks wander. The central axis of the gardens aligns with the centre of the Sala Terrena to establish the single line of balance within the asymmetrical walls of the site and the presumed chaos of the world.

Windows projecting from the roof display "inverted volutes". They are no longer used as masks, but as an optical play to exaggerate visual tension at the corners. The use of diamond shaped windows (indicating the geometry of the stair) also accents this expression of movement.

On the southern side of the garden is a wall 170 metres long and 8 metres high with a strange ornamentation of an unpleasant texture. Grey and hidden in the shade, figures of tormented faces and beasts appear in the scraggly surface of this massive cavernous cloth. Within the wall, there is the grotto and the private spa. The dark and menacing character of the wall is meant to be unattractive and disrupt the view from the Castle.

The rooms of the palace are interconnected to one another, each shows a unique feature. All are furnished with ornate marble and wood detailing: chandeliers, decorative walls or leather wallpapers, mirrors, crystals and painted ceilings, all emphasising Wallenstein's mythical power. An exceptional room is the Grand Hall within the main palace. Doors and stairs to this space are high and large enough to walk through on a horse. Painted ceilings and mirrors further exaggerate the spatial limits and the visual plays of perception. Another exceptional space is Wallenstein's private chapel, a hidden towering space within the Palace, rich in painting and gold reliefs, with a small window near its top facing towards Prague Castle.

Today, the palace is owned by the people and houses the Senate of the Czech Republic. It has been part of the National Museum and the Ministry of Education. The gardens are open to the public and serve as exhibition and concert grounds.

克拉姆 - 葛拉斯宫
Clam-Gallas Palace

地址：Husova 158/20, 110 00, Praha 1, Staré Město
交通：Tram 2, 14, 17, 18, 93 **Stop** Staroměstská（5 分钟步行）
建造时间：1713—1719 年
建筑师：约翰·伯恩哈德·菲舍尔·冯·埃拉赫（Johann Bernhard Fischer von Erlach）　　□ 巴洛克 / BAROQUE

克拉姆-葛拉斯宫位于皇家大道上，坐落于克莱蒙特图书馆、犹太区和老城区的交会处。作为当时维也纳哈布斯堡王朝的辖区，克拉姆-葛拉斯宫是省会布拉格的权力所在地。尽管建筑的正立面在狭窄的街道上不易被观察全貌，但它却是一座雄伟的建筑。宫殿建于1713—1719年，是波希米亚的首席执行官约翰·文策尔·冯·葛拉斯的府邸，由维也纳的皇家建筑师约翰·伯恩哈德·菲舍尔·冯·埃拉赫设计。这是一座纯维也纳巴洛克盛期宫殿建筑的典范，将权力、地位和复杂精巧表现得淋漓尽致。

这座18米高的三层建筑外观装饰呈现出一种波动的韵律感。宫殿没有开设中央纪念性入口，但在主立面的两边各设一个入口。每个入口由两对披着狮子皮的半神雕塑守护，极具力量感，展示了该建筑具有的统治和威严感，并代表了居住者至上的地位和对社会秩序的管控。建筑北墙有第三个入口，虽不太精致，但在空间上更具雕塑感。它面向市场，一个大露台位于上层并可以远眺。

与华伦斯坦宫不同，这幢建筑内部完全私密，只在某些场合对外开放。建筑内部到处都是精美的巴洛克风格的装饰、壁画和雕塑。与文化起源相关的希腊图案和古代神话主题作为文化资本和社会地位的象征被重新使用。建筑内部的巨大的楼梯绘满了罗马诸神的主题壁画；在内部的大庭院里，还有一个希腊海神特里顿的喷泉。喷泉、楼梯和正面的雕塑都由马赛厄斯·贝纳特·布劳恩创作。在众多的厅堂中，有一个特别华丽的用镜子和水晶灯装饰的大理石厅，莫扎特、贝多芬等都曾造访这里并演出。

20世纪后，克拉姆-葛拉斯家族因缺乏建筑维护资金，向外出租了一些房屋。弗兰兹·卡夫卡在完成法律学习后，就在这里开始实习工作。1945年第二次世界大战后，克拉姆-葛拉斯家族移居国外，宫殿被收归国有，由布拉格城市档案馆使用。宫殿每两年都会举行专注于巴洛克艺术的音乐、舞蹈、戏剧和嘉年华的综合性演出。

... placed with three entrances, at the crossroads between the Royal Way, the Clementinum library and the market squares.

It stands without compromise, a far greater building than any nearby. It makes an effort to demonstrate its broadly cultivated knowledge and to represent the primacy of the top Imperial official in Prague. A few generations after Wallenstein, the highest strata of the Crown were living in a much less tense environment under the Habsburg rule based in Vienna. Dozens of palaces were being constructed throughout the city to accommodate the new aristocracy.

As an extension of Vienna, the Clam-Gallas Palace is the seat of power in the now provincial city, Prague. It was built from 1713 to 1719, by Imperial architect Johann Bernhard Fischer von Erlach, who brought to it a High Baroque style, in a grand display of power, dominance and sophistication. After the site was selected in 1690, the land was gradually purchased. The architecture is as foreign, as its manifest and symbolic power — captured in its eight sculptural figures — the demigod men holding open the entrance gates, wrapped in their lion skins.

Rising high above the street to a height of 18 metres, it is a three-storey building. This was the house designed for Johann Wenzel von Gallas, the Chief Marshall of the Kingdom of Bohemia and international diplomat. In 1757, it was inherited by Kristian Filip of Clam. The Palace is positioned as the mediating figure between all those who would pass along this busy and intermixing point in the city. It faces the Royal Way to the south, the Jesuit Clementinum to the west, the merchants around Old Town Hall to the east, not far from the former walls of the Jewish Town to the north. At this crossing point between significant political players, the Palace subtly shifts and bends itself to manage the complexity around it.

The building displays a volatile but tempered ornamental expression. Along the narrow site of Husova Street, the facade is predominantly flat, but highly sculptural. Its main entrance points are approached from an oblique angle. There is no central monumental entry, but rather two peripheral entries, one at each corner. The extreme displacement and exaggeration of the points of entry from the street, in this case, are also a display of the political management of the social orders of the city. It was, after all, a house for a foreign diplomat, negotiating and re-directing the movements of the people through its control over the streets.

One's attention is pulled to the magnificently carved entry gates. Compared to the early figures of the Gothic — still absorbed in low relief within the engravings of the wall — the symbolic figures at this point developed a new message of mythic strength and power. Two such gates were placed, inviting, with equal authority, visitors from either side. The north wall had a third entry, less elaborate but more sculptural in space. It faced towards the markets, with a large terrace set back on the upper level, looking out from above.

Unlike in the Wallenstein Palace, the interior here was fully private and not on display. The Palace opened only for certain occasions. Porticos and sculptures adorn every surface with refined materials, craft and symbolism. Ancient mythology, associated with the origins of culture, were re-appropriated as a symbol of cultural capital and standing in society. There is a gigantic staircase in its interior, completed with painted frescoes of Roman gods. There is also a fountain of Triton the Greek god of the sea, within a large enclosed courtyard. Among the meeting rooms for events and guests and social spaces for diplomatic functions, there includes a vast marble ballroom, the Marble Hall, with mirrors and crystal chandeliers. Beethoven and Mozart performed here.

Today, the palace is owned by the City of Prague which lends it to the Prague City Archives. A multi-genre festival, focused on historically informed performances of Baroque music, dance, theatre and carnival takes place in the palace premises twice a year. It is planned to be renovated into a cultural public space to be open year round.

圣尼古拉斯教堂（小城）
St. Nicholas Church (Lesser Town)

地址：Malostranské náměstí, 118 00 Praha 1, Malá Strana
交通：Tram 5, 7, 11, 12, 15, 20, 22, 23, 41, 97 **Stop** Malostranské náměstí
（2 分钟步行）
建造时间：始建于 1673 年
建筑师：克里什托夫·丁森霍夫（Kryštof Dietzenhofer）、基利安·伊格纳克·丁森霍夫（Kilián Ignác Dietzenhofer）、安塞尔莫·卢拉戈（Anselmo Lurago)等

□ 巴洛克/ BAROQUE

位于小城广场的圣尼古拉斯教堂是中欧地区最华美的巴洛克建筑之一。1673年，场地上原有的一座哥特式教堂被拆除，这座巴洛克教堂开始建造。从1673—1752年，教堂的主要工程在这个广场狭窄的斜坡上持续近80年。建筑西立面和教堂中殿由克里什托夫·丁森霍夫在1703—1729年建造。1729—1751年，他的儿子基利安·伊格纳克·丁森霍夫将中央穹顶扩大，1751—1755年安塞尔莫·卢拉戈完成了钟塔的建造。1761—1770年乔安·克拉克完成了室内的壁画。教堂的建造在社会变迁中跨越了将近一个世纪。

由于修建教堂的一部分资金来自市政，因此市政便影响教堂的建造，而不是由教会单独主宰教堂的建设。最终二者达成妥协，产生了一个双核心的建筑结构：教堂的穹顶和城镇的塔楼并置，塔楼和穹顶均到了79米的高度。在这样一个建筑构图中，两个主导元素同时达到了既对立又和谐统一的状态。

伴随着不同阶段的建设，时代的演替在教堂上留下了独有的印记。尽管建筑体上许多适宜变化都是原生自发的，但设计从始至终是连贯的，甚至是一个内部有着相互关联的整体。作为一个巴洛克的艺术品，这是一个可以在更长时间内看到不同思想演变的很好案例。尽管圆顶教堂和塔楼这两个构筑元素是在不同时期以不同风格建造的，但它们被视为统一的建筑构图。塔楼和穹顶的高度相同，但塔楼看起来更高，这是教堂穹顶的起拱点较低所引起的视觉错觉。站在地面人的视角上看，塔楼的垂直性被夸大，看起来比教堂要高出好多。从查理大桥看过来，教堂的塔楼和穹顶成为布拉格城堡的前景，小城区最重要的空间标志。

教堂入口处，华丽的巴洛克风格的花岗石楼梯完美结合了起伏的地形，并融入了建筑起伏的立面。建筑整体的力度被群柱放大，这些群柱同时也打断了楼梯的整体连续性，制造出一些起伏波折。教堂的穹顶直径20米，高50米，是布拉格最高的建筑室内空间。教堂上最后建造完成的钟塔采用巴洛克的后期形式——洛可可风格，雕塑明显变得扁平化。它阶梯形的收分、层叠的壁柱和折角使其具有活泼动态的外观，并与穹顶形成对比。

建筑内部空间是整个巴洛克篇章的高潮。空间中每个独立部分在不同程度上按照巴洛克风格的可塑性进行建造。大理石的表面铺装，绚丽的彩绘天花，精致的建筑细部以及雕塑使得教堂内部空间的视觉表现极佳。整个建筑空间内部通过重叠的空心椭圆体结构获得的模糊性似乎产生了混响以及声音放大的效果，这与教堂内部安装的 4000 管的管风琴完美地契合，莫扎特曾于 1872 年在此进行《C 小调大弥撒》的首演。

如今，圣尼古拉斯教堂和市政钟楼与布拉格的天际线已经密不可分，体现出巴洛克风格建筑随岁月的动态变化以及张弛有度的风格特性。

...on the edge of the centre of the Lesser Town square, a dialogue between forces surges for the highest ground.

While returning to centre-stage and stability in the Counter-Reformation period, the Church could not supersede the towns in a society increasingly governed by the market and by people. The construction of St. Nicholas Church spanned nearly a century of changing dynamics and different visions. Negotiations involving the Church, merchants, the municipality, and noblemen (with competing allegiances to either Jesuit Catholicism or Hussite Reform), as well as three architects — father, son, and apprentice — resulted in a particularly unusual and hybridised Baroque Catholic church within the centre of the Lesser Town.

Part of the property beneath the church belonged to the town and the town therefore expected to influence what would be built. Rather than allow the square to be defined by the church alone, a compromise was reached. The dense and compact shape of the site required a massive vertical form, and the town required a tower, resulting in a two-part dominant structure. Within this composition, a unified, if contrasting, harmony was achieved through both dominant figures, tower and dome, reaching the same maximum height of 79 metres.

With each phase of construction and with each generation, however, the changing times made their marks on the building's appearance. Many adaptations were made from the original, yet the design remained a coherent, even interconnected whole. As a Baroque artefact, it is an opportunity to see the transformation of ideas over a longer period of time. The major construction works lasted more than 50 years, from

1673–1752 on a narrow, sloping site in the market square. An existing Gothic church on the site was demolished to begin construction of the foundations for a new Baroque church in 1673. The western facade and nave, facing away from the market, were built by Kryštof Dietzenhofer from 1703 to 1729. The central dome was enlarged by Kilián Ignác Dietzenhofer from 1729 to 1751, and the belfry tower completed by Anselmo Lurago from 1751 to 1755. Interior frescos were finished by Johann Kracker from 1761 to 1770.

The architectural expression of the domed church and tower belfry are treated as a single composition, despite the fact that the two were constructed at different times and with different expressions. While the tower and the dome have the same height, the belfry tower appears taller. Because the dome of the church begins lower than the belfry, the top of the stone wall ends at a lower height. This perception exaggerates the verticality of the tower, making it appear to extend a few floors higher than the church's dome. From the point of view from the Charles Bridge as well as from the south, the tower, like that of the Prague Castle is represented in the foreground. The tower as a civic symbol again appears to dominate the higher order of the Church.

The marble stairs at the opulent Baroque entrance blend the terrain and the undulating facade. The dynamism is amplified by the clustered columns which break the flow of the stairs, making waves. The dome is 20 meters wide and at 50 meters is the tallest interior space in Prague. The belfry tower, built last, is in the latest form of the late Baroque, in which the sculptural dynamics of the form are significantly flattened. Its stepped, horizontal rhythms playing against the dome, its layered pilasters and folded corners, give it the appearance of movement.

To varying degrees, each separate part of the entire structure was built with plasticity. This is perhaps best seen in the interior nave. The ambiguity achieved through overlapping void elliptical volumes appears to produce the perception of reverberation and acoustical amplification, appropriate for the 4000 tube organ installed within the church and played by Mozart with the premiere of his *Mass in C* in 1782. Elaborated with marble, illusionistic painted ceilings, golden details and sculptures, it is the high note of a long Baroque chord.

Today, St. Nicholas Church and the municipal belfry are inseparable from the skyline of Prague and embody the dynamic changes, flexibilities and tensions characteristic of the Baroque time.

斯特拉霍夫修道院及其图书馆
Strahov Monastery and Its Library

地址： Strahovské nádvoří 132/1, 118 00 Praha 1, Hradčany
交通： Tram 22 Stop Pohořelec（6 分钟步行）
建造时间： 1140—1797 年
建筑师： 安塞尔莫·卢拉戈（Anselmo Lurago）、伊尼亚齐奥·帕利亚迪（Ignazio Palliardi）

罗马风/ ROMANESQUE
哥特/ GOTHIC
文艺复兴/ RENAISSANCE
巴洛克/ BAROQUE
洛可可/ ROCOCO

布拉格城堡上方的斯特拉霍夫修道院，坐落在佩特任山上，俯视着山下的小城。1140年，弗拉迪斯拉夫二世为维护普雷蒙特雷修会的秩序修建了这座修道院。修道院内的圣母升天教堂为教宗所封的宗座圣殿。修道院图书馆在9—18世纪藏书有20多万册，包含宗教、医学、数学、哲学、地理、天文等主题，是波希米亚最具价值和保存最完好的古老图书馆，因精美的装修也被誉为欧洲最美的图书馆之一。

修道院在历史变迁中整合了各个时代的形式和思想，并组成更大、更复杂的整体。原始的罗马式修道院是方形院落布置，罗马式的巴西利卡有中殿和两个侧廊。巴洛克风格的重建定位了修道院的东翼，使其能够俯瞰城市，从而成为标志性建筑。1742—1758年，圣尼古拉斯教堂钟楼的建筑师安塞尔莫·卢拉戈对修道院的内外，包括塔楼进行了重建。

斯特拉霍夫图书馆毗邻大教堂入口，在场地中居主要位置。该图书馆作为原始修道院西翼的延伸并成为进入场地后看到的第一座建筑。与修道院的东翼一样，图书馆的立面是一个独立形象。巴洛克风格的主题和形式：如破碎的山墙，椭圆形和对柱，重新融入立面，但不再以空间破坏的方式；装饰物和人像的过度展示被拉平从而弱化了雕塑感。图书馆正立面的后面就是楼梯，但在立面上没有任何显现，显示了图书馆内外部之间概念和构造的分离。

图书馆有两个瑰丽的藏书大厅，两层高的哲学大厅于1782—1784年间由伊尼亚齐奥·帕利亚迪设计，是整个修道院图书馆内最大的厅，长32米，宽10米，高14米。其内壁可再次看到洛可可展平的雕塑，呈现出动态且柔美的形态。其室内精美绝伦的木装修由扬·拉霍夫于1794—1797年建造，富丽堂皇的顶面壁画由安东·莫尔伯奇绘制，描绘"人类对真正智慧的追寻"。同层走廊尽端，是更加古老的神学大厅，建于1671—1679年之间。天花略低，有着华丽的巴洛克拉毛装饰，以及描述"真理智慧"主题的彩绘涡卷。

斯特拉霍夫修道院是布拉格最古老、使用最久的修道院。修道院图书馆至今仍然是学习古代知识、科学和艺术的中心。该修道院仅在18世纪末和20世纪50年代初期短暂关闭。除了修道院、教堂和图书馆外，还拥有一个出名的啤酒作坊。

… on the upper recessed slopes of Petřín Hill, a sacred acropolis.

It was already established, centuries earlier, at this location above the city and above the Prague Castle. Within the fortifications of the city walls and overlooking the landscape at the gateway to Hradčany, the monastery retained its position and even strengthened it during the disruptive periods. It is still the central authority of knowledge, science and art, upon an elevated and protected ground, standing guard at the highest point in the city.

The Strahov Monastery's privileged position has not changed since its Romanesque foundations in 1140. In a hidden part of Petřín hill, it does not occupy a position that offers a fortified advantage, yet it controls the upper, western gates to the Castle. Due to its geographical and political autonomy, the morphological history of the Monastery reveals the transformations of an internal debate about self-representation free from public influence. Throughout the history of its accumulation and

revision, it adopted and re-integrated the forms and ideas of its times. It is a composite of multiple structures which create a larger, complex whole.

The original Romanesque monastery was a square-shaped cloister with an inner courtyard and a Romanesque basilica with a nave and two side aisles. The basilica was not meant to regard the city below. Baroque reconstructions, however, reoriented the eastern wing of the monastery so that it would stand upon the horizon overlooking the city, and thus become a dominant figure on the upper horizon. Other reconstructions to the basilica inside and outside, including the towers of the Church, were made in 1742–1758 by Anselmo Lurago, architect of the belfry at St. Nicholas Church. (While the Church may have been compromised by the tower in Lesser Town, at Strahov, it would stand proud.)

In a primary location, adjacent to the entry to the basilica, stands the enlarged Library. This Library (also called the Philosophical Hall) was added to the complex between 1782 and 1784, by Ignazio Palliardi, as an extension of the 17th century Theological Hall, at the western edge of the original cloister. The wooden library interior was built from 1794 to 1797, by Jan Lahofer, with its painted ceiling by Anton Maulbertsch depicting the "Journey of Mankind to wisdom" in 1794. It is the first building seen upon entering the grounds and the last phase of the Monastery's expansion. Of particular significance is the front facade. As with the Monastery's eastern wing, the Library facade was a separated, angled form which projected into the public space. The Library is a rectangular volume, 32 metres long, 10 metres wide and 14 metres tall, yet its main facade, in an almost surreal gesture, was conceived as a free-standing mask.

The front facade is angled to draw attention to the church entrance. But it also reveals a conceptual and tectonic separation between the exterior and the interior of the Library. Just behind this facade are stairs, which have no indication on the facade. Motifs and forms of the Baroque, such as broken pediments, ellipses, and column pairs, were reintegrated into the facade, but no longer in a spatially disruptive way. The extravagant displays of ornamentations and figures are flattened into less sculptural forms.

This shift towards the pictorial, flattened form makes use of the rhetoric and forms developed throughout the Baroque, however without the instabilities or risk of overt conflict and tension. The re-inscription of messages on the facade in the style of Imperial Roman civilisation (reading "to the benefit of religion, homeland and Zion") underscores a new interest in establishing a definitive social order and reflects the desire, at this moment in time, for stability, clarity and calm.

Within the walls of the two-level library can be seen again the Rococo flattening of the sculptural, in a dynamic but subdued form. The spatial volume is essentially cubic, despite its ornate and undulating walls and its curved, painted ceiling. As a representative space of peace and knowledge for the world, the walls of the Library reestablish order, by calming and flattening the tensions in a move towards a rational and stable balance. The historical moment of spatial rupture has been internalised and contained within the stillness and silence between the surface mask and what the mask conceals. The Library holds 200 000 publications including religious, medical, mathematical, philosophical, geographical, astronomical and other themes.

Strahov Monastery is the oldest, continuously operating monastery in Prague, closed only briefly at the end of the 18th century and the early 1950s. In addition to the monastery, church and library, it also houses a brewery.

古典复兴时期

?—1781—?

这个时期也许起始于玛丽亚·特蕾莎女王1742年开始统治波希米亚,或是起始于1781年废除农奴制及波希米亚城市的工业化,又或是起始于1848年的民众起义,或又在1881年,随着城市的扩张而起,抑或是由于1874年布拉格城墙的拆毁,或1834年第一列火车的到来。可以肯定的是,到了1918年第一共和国成立,这个时代在多重异质因素、技术发展和社会冲突的影响下发生了彻底转变。

农奴制的废除使民众有了更广泛的职业选择,波希米亚成为哈布斯堡帝国的一个重要工业区。工业化使人群不断涌入城市,加速了布拉格城市化进程。同时在这个时期,捷克民族意识开始崛起,民族独立的努力最终在19世纪中的布拉格国家大剧院建设上体现出来。捷克语的地位不断增强,19世纪90年代,德语名称从路标上消失了,捷克语成为布拉格官员的行政语言。

这座城市的未来是由生产市民文化的公民自身创造的。因此开启了城市中一种新的社会景观:关于城市自身的内部讨论,每一个人都追求进步的理念。在中产阶级和受过教育的商人支持下,布拉格出现了一种新的、变革性的,关于自我表征、意义和未来的话语,这种话语建立在重建过去形象的基础上。

这是一个有着不同表达方式的复杂时期,提供了大量的"主题"和"风格"可供选择:一个无尽的符号目录,强调对称、清晰和稳固。这些风格可以被用来表达投资者的社会价值,也可以用来反映哈布斯堡帝国的范围和多样性。在这个时期,个人品位的社会价值与文化的象征价值融合在了一起。

在此处所定义的古典复兴时期中,风格之间会相互交叠,就像当时各种冲突的观念一样都处于一个不断变化的状态,并以自己的形式表现出来。但从一种风格到另一种风格都没有明显突破。

随着科学时代的到来和新兴学科的出现,人们对自然和历史进行了重新评价,通过新技术和新思想产生了一种对社会的重新表述,并将其重新应用到文明形象中。此时,人们能够用理性和判定来抽象外部世界,选择"历史上最好的范例"来表达当时的理想。

古典复兴时期的布拉格
The Revivalist City

新古典主义（1795—1830—？）

随着晚期巴洛克风格逐渐转变为洛可可风格的不那么易变的表达方式，由改革和反改革所激发出来的活力释放为一种更加稳定的秩序。巴洛克风格的图案和主题被吸收到扁平的表面，并与古代"古典真理"的意象和阐释相结合。这种风格显示出一种兴趣，即将过去的符号作为理想、稳定、无破坏性、中性的建筑形式的普遍（和"安全"）表达来引用。布拉格19世纪末和20世纪初的大多数建筑都以不同的方式展现了这些新古典主义原则。

过渡装饰
Ornamental Excesses

神话的自我再现
Mythical Self-Representation

新哥特式（1830—1880）

科学和工业的兴起标志着对哥特时代工程技术兴趣（其建构上的成就）的回归。与历史主义的哥特复兴不同，在新哥特时期，哥特建筑的结构原则被现代化。尖券拱和垂直空间的建筑意象被重新应用在新的技术语言和新的工业材料——铁和玻璃中。

新文艺复兴（1850—1900）

在将工业城市作为有机体建立起来后，布拉格开始被视为一种具有文化意义的人造物。在多起民众起义的背景下，城墙被拆除，城市的象征形象被重新定义，一个有着多样思想和日益独立的新的现代化国家出现。历史保护意识与布拉格的民族复兴相结合：一种"原初精神"的"重生"。伴随着新兴财富和公民社会的独立性，随之出现了相应的市民当局。这二者虽不同，但都通过重新召唤波希米亚王国的黄金时代，来提升城市的斯拉夫遗产。

个人主义
Individualism

赞助人和铭文
Patronage and Inscriptions

重组古代主题
Recombined Ancient Motifs

理想化的象征主义
Idealised Symbolism

新艺术运动（1890—1912）

作为一种新的，更适合表达"自由"的形式风格，新艺术运动在充满活力和变革的 20 世纪开端出现。捷克建筑师从巴黎及维也纳新艺术运动（分离派）的流动线条和对自然的赞美中获得启发，开启了布拉格的新艺术运动。这种建筑表面绘画式、装饰性的艺术目的是将新古典主义再次定义为"新事物"，将其定义为从过去到未来，从一个老派的社会象征和符号到当今技术和价值的展示。如今，布拉格可能是欧洲观赏新艺术风格建筑最佳的城市之一。

……所有这些古典复兴时代的风格都将在这段时期里，在不同建筑中展现出来。

12. 城邦剧院
The Estates Theatre

13. 金斯基花园
Kinský Folly and Garden

14. 马萨里克火车站
Masaryk Railway Station

15. 国家大剧院
National Theatre

16. 鲁道夫音乐厅
Rudolfinum

17. 布拉格中央火车站
Prague Main Train Station

18. 市民会馆
Municipal House

THE REVIVALIST TIME

?–1781–?

... perhaps it began with Queen Maria Theresa's influence on Bohemia in 1742, or in 1781 with the abolition of serfdom and the rise of industrial manufacturing in cities. Or later, in 1848, with popular uprisings. Perhaps again in 1881 with the expanded urban area, or perhaps, with the 1874 demolition of the city wall or the 1834 arrival of the first train. Certainly, by 1918, the period will have found itself fully transformed by the impact of multiple, divergent influences, technologies and conflicts.

From doubt, the awakening of change, comes expectation as the generator of more changes. This process of transformation also implied a process of perpetual contradiction. For this reason, the years before and after the 19th century are viewed as part of the dynamic variations of a volatile time between the Baroque and the Modern. During what we call the Revivalist time, styles overlap, as the many conflicting ideas of the time, all in a state of flux settled into their own forms. There is really no clear break from one to the next.

It was a complex period of many different expressions, which offered a vast range of "motifs" and "styles" to choose from: an almost endless catalog of symbols emphasising symmetry, clarity and stability. These styles could be employed to express a social value of an investor and also mirrored the range and diversity within the Empire. During this time, the social value of personal taste merged with the symbolic value of culture.

The city's future came to be initiated by private citizens offering civic and cultural content. This opened a new social landscape within the city — engaged in an internal debate about itself — each in pursuit of its own idea of progress. In Prague, with the support of middle classes and educated private businessmen, there emerged a new, transformative and transforming discourse about self-representation, meaning and the future, based upon the reconstructed images of the past.

With the scientific era, and emerging disciplines, came a re-evaluating of nature and history. With it, came a social re-formulation through new technologies and ideas, and its re-application into an imagery of civilisation. Human knowledge, capable of abstracting the world using reason and choice, selected "the best examples from history" to represent the ideals of the present.

Neo-Classicism (1795–1830–?)

As the late Baroque resolved into the less volatile expression of the Rococo, the dynamism which had been unleashed by the Reformation and Counter-Reformation was relaxed into a more stable order. The patterns and the motifs of the Baroque, however, were absorbed into its flattened surfaces, and were combined with the imagery and interpretations of ancient "classical truths". This style displays an interest in referencing symbols of the distant past as universal (and "safe") expressions of an ideal and stable, non-disruptive and neutral, architectural form. In various manners upon their facades, most of the late 19th and early 20th century buildings in Prague display these Neo-Classical principles.

Neo-Gothic (1830–1880)
The rise of the sciences and industrialisation marked a return of interest to the engineering of the Gothic time — for its achievements in construction. Unlike the historicist Gothic revival which focused on Gothic imagery, in the Neo-Gothic, the structural principles of Gothic constructions were modernised. Pointed arches and spaces of verticality were re-appropriated into a new language of technologies and industrial materials: iron and glass.

Neo-Renaissance (1850–1900)
After having established the industrial city as a mechanism or organism, Prague began to be operated upon as an artefact of cultural meaning. The city wall was removed and the city's symbolic image redefined. This was taking place in the context of multiple popular uprisings, and the beginnings of a new modernising state where there was a diversity of ideas and an increasing independence of the people. Conservation and historical consciousness were combined with a National Revival in Prague; a "rebirth" of the "original spirit". There was a corresponding emergence of the authority of the Municipality, with the emerging wealth and independence of a civil society. While neither was a homogenous entity, both advanced the Slavic heritage of the city by re-calling the Golden Age of the Bohemian Crown.

Art Nouveau (or Secese, Secession, 1890–1912)
Art Nouveau emerged as a new, more appropriate form of "liberated" expression for the dynamic and evolutionary time. It was a pictorial, ornamental and decorative architecture of the surface. Its aim was to reorient the Neo-Classical establishment from the past to the future, from the symbolism and references of an "old fashioned" society — with an expressive breakaway, a secession, towards the celebration of technologies and values of the present day — as "something new".

... All of these styles of the Revivalist time will be noted as they manifest themselves in different buildings.

城邦剧院
The Estates Theatre

地址：Železná, 110 00, Praha 1, Staré Město
交通：Metro A, B **Stop** Můstek（4 分钟步行）
建造时间：1781—1783 年
建筑师：安东·哈芬莱克尔 (Anton Haffenecker)

☐ 新古典主义/ NEO-CLASSICAL
■ 复兴主义/ REVIVALISM
■ 洛可可/ ROCOCO

城邦剧院是一座捷克民族音乐的圣殿。18 世纪后期，音乐和文化被带入布拉格的城市日常生活，并利用新古典主义即古希腊罗马的建筑语言来表达公民社会和文化进步。剧院建于 1781—1783 年间，由诺斯蒂克·里内克伯爵资助，安东·哈芬莱克尔设计。剧院建成后成为布拉格第一座新古典主义建筑。

剧院位于老城水果市场的中心，独立在广场中央，四个立面朝向四个完全不同的城市空间。其主要入口与泽勒兹纳街对齐并与老城广场连通。东立面朝向哥特时代的水果市场，南侧可直接通往护城河街。在剧院阳台以及（邻近）查理大学的哥特式建筑片段之间亲密对话中，老城的肌理和尺度也在北侧展开。西立面和门廊是城市和剧场之间的过渡空间，也是巴洛克和洛可可风格之间的时间过渡，它的形式融合了罗马式神庙和波希米亚的塔楼。两对科林斯柱让入口显得雄壮有力，与壁柱一起，形成了建筑三段式的主体部分。

剧院内部用精致材料和彩绘装饰。大厅内装修采用蓝、白、金三色。栏板花饰、线角和柱头为石膏花饰镀金。剧场建成时用蜡烛照明，阳台栏杆的浮雕表面会将这种水晶灯的烛光照射反射到空间中。虽然剧院已经过现代化改造，但整个建筑包括演出大厅都还保持着原貌。

剧院可容纳听众 1200 座，在独特的密实垂直空间中，五层半圆形包厢面向突出的舞台和开放的内陷式乐团区域。紧凑而垂直的包厢布局让观众感觉即使从顶层包厢也能接近演出舞台。但在此后这不再是剧场空间常用形式。

为了改进消防安全和扩建剧院，在 19 世纪末和 20 世纪初对建筑立面进行了改动，细致的修缮仿佛就在执行原设计。1882 年，为安全考虑在建筑侧面增加了铸铁阳台和楼梯，并将外立面窗户细分与内部的出口走廊相匹配。在 1920 年，建筑再次被改建，插入一个钢结构体和一个附属构筑，从地下连接到街对面的科洛弗拉特宫。20 世纪初增加了建筑背立面的壁柱、双壁柱以及科林斯柱，它们的曲线造型与西柱廊相呼应。

作为布拉格现存最古老的剧院，捷克文化史上的很多重要时刻都在此发生：1787 年莫扎特歌剧《唐璜》的首演；1820 年捷克语舞台剧的首演；1827 年捷克语歌剧的首演；1834 年泰尔的喜剧《不生气，不吵架》的首演（首演中的歌曲《哪里是我家？》后来成为捷克国歌）；1983 年美国籍捷克裔导演米洛斯·福尔曼回到故乡，在此拍摄了奥斯卡获奖影片《莫扎特传》等。

如今，作为国家剧院的分部，人们可以在这里欣赏包括《唐璜》在内的戏剧、歌剧和芭蕾舞剧等各类演出。

...at the centre of the busy fruit market in the Old Town, a temple for the Music of a Nation.

Music and culture were brought into the daily life of the city by adopting the Roman iconography of the temple to express a new era of the cultural advancement of civil society. Funded by Count Nostic-Rieneck, the Estates Theatre was built between 1781 and 1783. Significant moments in the history of culture took place here. For instance, in 1787, Mozart's *Don Giovanni*, in 1820, the first Czech theatre performance, in 1827, the first Czech Opera, in 1834, the premiere of Týl's play *Fidlovačka* and its song *Kde Domov Můj?*, which later became the Czech national anthem, and the 1983 filming of Miloš Forman's award-winning *Amadeus*.

The Estates Theatre is free-standing upon the square with four completely different spaces on its four sides. Its large rectilinear volume is defined primarily by the western portico, with its main entry aligned with Železná Street and linked to the Old Town Square. The east facade faces the Gothic-era fruit market, Ovocný Trh, and the south side has a direct link to the first civic boulevard on Na Příkopě. The fabric and scale of the Old Town are also engaged on the north side within the intimate dialogue between the theatre's balcony and the Gothic fragment of Charles University.

The west facade and portico is a transitional moment between the city and the stage within. It is also on the temporal edge between the Baroque and the Rococo. As a sacred point within the city fabric, its form is a blend of the Roman church and temple, with the Bohemian tower and house. The portico entrance—not quite sphere or tower — stands guard over the civic temple held within. From the exterior there is no indication of the theatre hall inside. Yet its unique appearance indicates a temple of cultural significance. The description PATRAE E MUSIS (inscribed "a la Romane") was an announcement for a new culture — not the banner of a shop. While its window proportions have changed, the original and subtly curving projection from the building mass remains.

These only curved walls of the building are not intended to disrupt order, but rather to make a distinction between the main rectilinear volume and the figural, almost human-scale folly at its gate. The break of geometries at the meeting of these two volumes is blurred with the trompe l'oeil pilaster, whose fold in space makes it appears as a three-quarter column. Along the rising curve is a sequence moving from flat pilaster, to a (false) three-quarter column, to a pair of Corinthian columns in the round. The portico's plasticity is contrasted sharply with its flattened rectilinear base; a contrast between silence and sound.

Changes were made to the facade in the late 19th and early 20th centuries, as part of fire safety improvements and theatre expansion. They were carefully aligned to appear as if part of the original plan. In 1882, cast-iron balconies and exit stairs were added and the facade windows were halved in height to match the subdivisions within. In 1920, it was again rebuilt with an inserted steel structure and a new annex added with an underground connection to the Kolowrat palace across the street. The back facade, added in the early 20th century, echoes the Western portico: with

an implied curvature from pilaster to double pilaster to Corinthian column.

Within the theatre, a compact and vertical arrangement of balconies gives the audience a feeling of being close to the performance even from the top balcony level. Its unique type of compressed, vertical space has five levels of semi-circular balconies facing the inset stage and the open, sunken orchestra pit, and is no longer a commonly used form of theatre space.

The Rococo interior ornamentation is subdued, yet still made of refined materials and painted decorations. It was meant to be a performance space, rather than a performance in itself. As it was once lit with candles, the sculpted relief surfaces of the balcony balustrades were made to reflect this illumination into the space. While it has been modernised, the performance hall and the building as a whole remains mostly original in character.

The Estates Theatre has been operating continuously since its establishment, without significant interruption. One can still attend performances (including *Don Giovanni*) there today.

金斯基花园
Kinský Folly and Garden

地址：Petřínské sady 98, Praha 5, Smíchov
交通：Tram 1, 2, 6, 9, 12, 15, 20, 98, 99 **Stop** Švandovo divadlo
（5 分钟步行），Cableway from Újezd **Stop** Petřín（20 分钟步行）
建造时间：1825—1831 年
建筑师：海因里希 · 科赫 (Heinrich Koch)

☐ 新古典主义/ NEO-CLASSICAL

在小城外佩特任山的南坡上，金斯基花园通过"英式花园"的景观确立了自然在城市中的地位。它既置身于大自然之中，也坐落于私人宅地里，同时有着公共和私密的两个层面。在浪漫主义时代的高潮中，它代表了在艺术中追寻自然，以及希腊罗马哲学的回归。

1781年农奴制的废除和工业化的开始带来了建筑业的热潮，更多的人从农村涌向城市，使城市不断扩张。在1784年，原本分离的镇区并入了布拉格。金斯基花园在这个时期出现似乎是阻止了建筑地产大开发的势头，并由此保留了城市中重要的绿地。在空间上和时间上，它处于历史城镇和新城镇发展的交汇处。这类庄园曾经位于城市的郊野，如今由于城市扩展却成为城市中心的绿地。

1825年，鲁道夫·金斯基伯爵开始建造这个花园，与其说是必要，不如说是象征：用浪漫主义新古典的手法对自然空间的前瞻性保护。在1827—1831年间，花园中的别墅由维也纳建筑师海因里希·科赫设计。在1831年，建筑南墙上又增加了温室。

这是一个建在斜坡上的两层建筑，这样的布置使得金斯基宫可以同时拥有分别面对城市和自然的两个朝向。建筑被放置在重构的景观当中，并与自然要

素相融合。建筑前宽阔的草坪提供了非常开阔的视野。西侧地势较高，地面几乎与上层的阳台平齐。东侧地势较低，可以看到其完整的两层形象。建筑将自身与起伏的土地融为一体，成为这片土地的标志和连接物。

金斯基宫有着整洁鲜明的外表，除了凸起的入口门廊外，没有任何多余的装饰。支撑前阳台的四个独立柱形成了一个"隐忍谦逊"的柱廊，代表着罗马秩序中神殿的正面。当人们从正门进入建筑时，造型质朴的罗马柱仿佛从土地中生长出来一般。

作为金斯基宫的一部分，这个占地22公顷的开阔花园曾经是森林和葡萄园，现在则改造为人工景观，在花园里还有一个木制教堂。金斯基花园以一个值得被赞赏和学习的布局，展示和歌颂了自然的多样性。

1901年，为了防止其被拆分拍卖，金斯基花园被收归为政府所有。在1902年，它成为国家人类学收藏馆，陈列着一系列的古老家具和服饰。

…just outside the Lesser Town, on the southern slopes of Petřín Hill, a romantic synthesis of city and country, science and nature.

It re-establishes the primacy of the land within the city, in an "English garden" landscape. The building is firmly placed within a dialogue of the topography and the levels of the ground. The Kinský Folly has both a public and a private face. It is set in a natural background, and is located within its own private gardens. At the height of the Romantic Era of its time, it represented a return to nature in the arts, and to the Greco-Roman philosophies.

With the 1781 abolition of serfdom and the beginnings of industrialisation, a building boom occurred, bringing more people from the countryside into the city. In 1784, the separate town districts merged into the Imperial and Royal capital of Prague. The Kinský Gardens likely prevented other building developments from taking place at this time, and preserved important green space within the image of the city.

The gardens were built by Count Rudolf Kinský in 1825 — not so much for necessity as for symbolism — as a forward looking preservation of natural space in a romanticised Neo-Classical style. Between 1827 and 1831, the villa was constructed by a Viennese architect, Heinrich Koch and in 1831, the Greenhouse was added on the southern wall. This type of estate is unusual for a city centre, and was originally located just outside the incorporated city.

Built at the base of a steep slope, it is a two levelled building with two separate entrances, one on each level. It finds itself on the cusp within two natural folds: at the edge of its physical landscape and at the edge of its pre-industrial time. It was built physically and temporally on the convergence between the historical city and the new towns. Perpendicular to the slope of the hill, the Kinský Palace is placed so that both city and nature can be seen from each side.

The building is built into the reconstructed landscape and internalises it. On the western side is the upper level, and on the eastern side is the lower level and the elevated balcony. The long approach through the front lawn offers a partially screened view of the entrance. The "natural" ground level aligns with the upper-level balcony. At this moment, a path on the left moves to the lower ground, while one on the right moves to the upper ground. The building —as the symbolic and physical connection in the land — joins the splitting ground within its built form.

Placed within its own private landscape, the solitary, rational figure stands as a temple on a hill, with a degree of nostalgia, separate yet oriented towards the city folk. It has a puritan, stark and even bare exterior, with minimal ornamentation, apart from its raised entry portico. The four free-standing columns which support the balcony are a stoic and humble colonnade and represent the symbolic temple front of the Roman order. As one approaches the building from the main entrance, the rusticated Roman roots (the columns) appear to rise out from the ground.

The vast garden, comprised of 22 hectares, was once forests and vineyards. The gardens were planted as part of the Kinský estate to showcase and celebrate the variety in nature in an arrangement meant to be appreciated and reflected upon. In the gardens there is also a 17th century wooden church, the Church of Saint Michael, which was transported from the eastern edge of Czechoslovakia (now Ukrainian land) in 1929 as a symbol of "bridging the lands". In 2020, it was destroyed in a fire, but it will be reconstructed according to the original details.

In 1901, the property came under Municipal ownership, preventing its subdivision, and in 1902 it became the exhibition space of the National Gallery of Ethnography, displaying a collection of historical furnishings and costumes.

马萨里克火车站
Masaryk Railway Station

地址： Havlíčkova 2, 110 00 Praha 1, Nové Město
交通： Tram 3, 6, 7, 11, 14, 15, 24, 26, 37, 41, 91, 92, 94, 96 **Stop** Masarykovo nádraží
　　　Metro B, C **Stop** Florenc（9 分钟步行），Metro B **Stop** Náměstí Republiky
　　　（3 分钟步行）
建造时间： 1841—1866 年
建筑师： 安东尼·云林（Antonín Jüngling）
结构工程师： 扬·佩纳 (Jan Perner)

☐ 新哥特/ NEO-GOTHIC
■ 复兴主义/ REVIVALISM

马萨里克火车站位于共和广场以东，主要建筑建于1841—1845年间，其建造时间比著名的巴黎北站和东站都早，也是布拉格最早的蒸汽机车火车站。作为维也纳与德累斯顿交通线上重要的枢纽，马萨里克火车站通过铁路将布拉格连通到更大范围的哈布斯堡帝国。这个项目不仅满足了城市不断发展和社会日益增长的交通需求，同时也变成那个时代新的技术和文化设施，标志着波希米亚社会在19世纪工业技术方面取得的成就。

设计火车站的建筑师为安东尼·云林，他还设计了捷克境内布尔诺、奥洛穆茨等主要城市的中央火车站。最初火车站只建造了两组新哥特风格的建筑，具有不同的内部特征，分别作为出发和到达厅。巨大的出发大厅在希伯尼斯卡街，大厅上有两座塔楼，塔楼上设有一个大钟。方形的到达厅有一个内院，并设有员工宿舍，呈现出一个城市小街区的模样。两栋建筑采用了相同的灰泥粉饰外墙。

火车站场地功能划分合理，到达和出发分开并在之间留下了空间，基础设施规划体现了城市作为一种旨在组织和高效运行的运动机制的科学概念。那时，火车旅行的豪华被描绘为城市生活的一部分，沿着希伯尼斯卡街被展现出来。

人们可以穿过老城门直接到火车始发大厅，月台上的火车清晰可见，或者可以从西面到达厅乘坐马车绕过工人住区进入城市。货站和货厅则隐藏在城墙后面。火车站建成了，但城墙依然存在，显示了社会结构的分离和社会阶层之间的实体及象征性的差异。

1862年，火车站主要等候空间安装了6米高的锯齿形顶棚。铸铁柱上架设木屋架及采光玻璃顶棚，将火车站所有部分连接在一起。建造尽管使用了工业时代的铸铁材料，但也结合了波希米亚外围地区的木结构技术。1869年，出发和到达之间的拐角处内开设了一座咖啡馆及餐厅，形成了一个新的社交空间，这也促进了不同阶层的融合。这种新的社会交往方式也开始从火车站蔓延开来。

该车站于1918年被以捷克斯洛伐克第一任总统托马斯·加里格·马萨里克的名字重新命名。1950年修复了出发厅

室内在第二次世界大战期间受到破坏的部分。20世纪80年代对地铁建设中造成的建筑改变也进行了修缮,恢复到车站最初的面目。在2014—2018年间,又修复了候车大厅的玻璃立面和钢木屋顶。

火车站在20世纪70年代和21世纪初曾被考虑废弃,但幸免于难。近期,一条连接火车站和机场的高速铁路在计划中,扎哈·哈迪德建筑事务所在火车站东北方向设计了一个全新的大型城市综合体,将商业、火车、汽车及地铁等整合成一个新的交通枢纽。这个象征性的城市门户将再次重新连接周围环境,并形成新的社会城市景观。

… at the gateway between the city centre and the Empire, a split between two sides.

A symbolic new gateway linked the city by rail to the greater Imperial landscape. This municipal project responded to the growing city and to the growing demands of civil society. It was a piece of technical infrastructure that also initiated a new cultural infrastructure.

The terminal station follows the important historical path which links the centre to Vítkov Hill outside the city. The symbolism of the station was that it had industrialised this linkage, bringing the world outside the city walls into the city centre along its major axis. Six new gates were built through the still-standing city wall between 1841 and 1845, by the time the first train arrived into the station (The position of the city wall is now marked by the raised highway in its place).

To express the departure and arrival halls, at first only two buildings were built. Designed in a Neo-Gothic style, these two main buildings have different internal characters. The massive, linear departure hall had a clock on its two towers along Hybernská Street. This signified a point of industry and technological achievement on the ancient axis between the Powder Tower gate of the Old Town and the slopes of nature in the background. The square arrival hall, with its internal courtyard, assumed the form of a small city block, and included housing for workers.

The organised site plan which separated arrival and departure spaces also left a space between them which was later filled. The luxury of travel was represented as a part of city life, with its once exposed view along Hybernská Street. The infrastructure plan demonstrated the scientific conception of the city as a mechanism of movements, designed to be organised and efficiently run. One could leave the city through the Old Town gate directly towards the departure terminal, with the trains in the station visible. And it was possible to enter the city by horse-carriage immediately from the western arrival hall. Yet the working stations and industrial halls were hidden behind the city walls.

The station provided some housing and amenities for rail workers, but the social hierarchy of the bourgeois society of the time was embedded within the two halves of the civic building. While post offices, police stations and workers' apartments were included on the side of the city, the storage and service buildings were placed out of sight. That the train station was built, yet the city walls remained, showed the division within the social fabric, and demonstrated the physical and symbolic difference between the social classes.

After the 1848 uprisings, the concept of infrastructure was applied into the cultural life of the city. The construction of a restaurant and café, in 1869, on the corner between the departure and arrival station, established a new "Café society" as a centre of social interaction and a place-in-between for the mixing of classes. This new cultural space extended up and down the streets surrounding the train station.

In 1862, a cast-iron, wood and glass canopy was installed to cover the main waiting space between the three buildings, and to join together all the parts of the ensemble. It recalls the ancient form of a temple, albeit with an industrial material palette. It combines the timber construction techniques found throughout the outer regions of Bohemia with new construction technologies. This synthesis of the hinterlands, the industrial city and ideal symbols, capped the main waiting hall under a single roof covering for all, upon 6-metre-tall cast-iron columns.

The station was renamed in 1918, after Tomáš Garrigue Masaryk, the first president of Czechoslovakia. On May 8, 1945, during the Prague uprising, the Masaryk Station was the site of a massacre of civilians by fleeing German soldiers. The departure hall interior was remodelled in 1950 after these damages from the war. After changes made during the 1980s with construction of the metro, renovation of the Station began again to return it to its original form. More recently, from 2014 to 2018, the glass facades and the wood and iron roof above the waiting hall were restored.

The station's complete demolition was considered in the 1970s and 2000s, but this has been averted. In the near future, a high speed train link to the airport is planned, and new buildings by Zaha Hadid Architects will be built to the north and east of the station. They will link into the nearby bus and metro stations as a planned transit, office and shopping hub. This symbolic city gateway will again physically reconnect its surroundings and join social groups together with new pathways above the railroad tracks and below the highway overpass.

国家大剧院
National Theatre

地址： Národní 2, 110 00 Praha 1, Nové Město
交通： Tram 1, 2, 9, 17, 18, 22, 25, 41, 93, 97, 98, 99 **Stop** Národní divadlo
（3 分钟步行）
建造时间： 1868—1883 年
建筑师： 约瑟夫·齐克特（Josef Zitek）、约瑟夫·舒尔茨（Josef Schulz）

□ 新文艺复兴／ NEO-RENAISSANCE
■ 新哥特／ NEO-GOTHIC

国家大剧院位于老城的城墙遗址和伏尔塔瓦河的交会处，主立面对着民族大街，侧立面沿着伏尔塔瓦河展开，庞大的体量醒目地占据了整个街区。作为布拉格城市结构和民族文化的重要地标，国家大剧院凝聚了民族希望和国家认同，是捷克民族复兴和国家身份建构的象征。

19世纪末，捷克民族复兴运动高涨，新兴的民族主义运动期望建立一个斯拉夫文化的国家大厦，意在复兴布拉格的斯拉夫民族传统。虽然受部分哈布斯堡统治阶层的反对，但是剧院建设收到了来自捷克境内外各个阶层的捐款，包括建筑奠基石都具有象征意义地来自捷克的各个地区。

国家大剧院由约瑟夫·齐克特设计，1868年开始奠基并进行建设，于1881年6月11日首次开放，以欢迎鲁道夫王储的访问。建筑揭幕典礼是捷克民族音乐的奠基人贝德里赫·斯美塔那新作《里布舍》（讲述了一位公主创建布拉格城的传奇）的首演。但两个月后，建筑意外被火灾烧毁。约瑟夫·齐克特的学生约瑟夫·舒尔茨继续了项目重建，1883年11月18日，该剧院再次以《里布舍》的演出重新开放。

大剧院扮演着城市重要的景观元素。剧院与伏尔塔瓦河和布拉格城堡形成特殊的联系，并成为民族大街的文化高地。在剧院公共门廊露台可以纵览民族大街的街景，并可远眺布拉格城堡的景致，剧院朝北的门廊也融入了城市公共空间，框选出河对岸佩特任山景色和小城的城墙遗迹。

大剧院采用了非传统的新文艺复兴风格，适应了当地的形式和其场地的特殊性。一个显著特征是外立面参考了城市原始的哥特式语言，采用了完全裸露的石块。剧院的门廊突出在城市的街道上，屋顶上的金色皇冠也与布拉格哥特遗产的塔楼相呼应。建筑立面也并非典型的新文艺复兴风格的平坦墙面，而是依据基地的几何形状赋予建筑物棱柱体块感，每个立面都有重点地形成了自己的表达形式。这些概念在20世纪70年代末国家大剧院的扩建项目新舞台的设计中得到了发展。

建筑内外部的装饰无论雕塑、雕刻

还是壁画,均出自捷克著名艺术家之手,华丽精美且极富表现力。剧场内金碧辉煌,以金色和红色为主色调,装饰细部表现着捷克的国家精神和意象。国家大剧院在火灾重建后就采用了电力照明,是欧洲最早引入电力照明的剧院之一。由于当时采用了非常先进的建筑设备技术手段,因此剧院在百年后无需太多现代化更新。

今天,国家大剧院仍是布拉格最主要的文化地标,常年举办戏剧、歌剧和芭蕾舞演出,为国家重要的文化活动服务。

…at the meeting of the Old Town wall and the Vltava river, a symbolic cornerstone and a new foundation of a resurgent Slavic city.

The emerging nationalist movement sought to establish a national house for Slavic culture, and intended to symbolise the rebirth of Slavic heritage from within the centre of Prague. Though opposed by the institutionalised upper class establishment, donations from an emerging middle class of industrialists were collected from throughout the lands. Even stones of the building's foundation were brought from all the surrounding regions.

While a freestanding landmark in the city, occupying an entire block of its own, it was also a part of a city block structure. It was built within the context of the more traditional Neo-Renaissance buildings along Národní Street, such as the Czech Savings Bank and Lažanský Palace by Vojtěch Ullmann, seen across the street. It was designed, however, in a different way. The traditional, flat, Neo-Renaissance style was adapted to the irregular particularities of its site. One notable characteristic difference is the fully exposed stone, a reference to the material of the city's Gothic language.

The National Theatre was built along the structural imprint of the Old Town Wall and the new river embankment. It established the cultural institution as a symbol which was inseparable from the structure of the city. In 1851, it was originally planned for the bottom of Wenceslas Square (Václavské náměstí), as the physical centre point of the city. At this time, Narodní had already been a city boulevard for 100 years, but with the construction of the embankment, the site was moved to the river. A temporary theatre, built by Ullmann in 1862, was originally placed on the southern end of this site. The temporary theatre was incorporated into the whole composition, and from the river the difference of the building's two parts is visible. In 1868 the construction work was commenced by architect Josef Zítek, and another architect Josef Schulz resumed the continued project after the 1881 fire which destroyed the roof and the stage, until its completion in 1883.

Rather than the idealised flat wall of the typical Neo-Renaissance buildings, at the National Theatre, the geometry of the site is used to give a prismatic form to the building, giving different emphasis to different facades. Each facade has its own sculpted expression, a concept to be extended in the Theatre's expansion in the late 1970s. After centuries of representative facades, the National Theatre stripped the stucco off "the Neo-Classicist mask" to expose the natural stone beneath. The golden crown on the roof also echoed the towers of the city's Gothic heritage, marking itself as a civic landmark, and its street level portico projects into the public space.

The building acts as a framing element of the city. The portico terrace for the public provides a view down the prominent street, as well as a direct view to the castle. Along Národní, with its particular relation to the river and the Castle, it marks the climatic end of the civic promenade. Seen from this street, the north-facing portico frames Petřín Hill and the remnants of the Lesser Town Wall (the Hunger wall) across the river.

The expressive sculptural quality of the building is also found in the qualities of the exterior and interior ornamentation. Finely carved stones, refined and expressively painted ceilings and walls decorate the interior. The theatre space is embellished with red and gold, artwork and details celebrating national symbols and narratives. The highly advanced theatre technology of the time, did not need to be modernised for another 100 years.

Today, the National Theatre is still hosting performances and important cultural and national events.

鲁道夫音乐厅
Rudolfinum

地址：Jan Palach náměstí 12, 110 00 Praha 1, Josefov
交通：Metro A **Stop** Staroměstská（3 分钟步行），Tram 2, 17, 18
　　　Stop Staroměstská（2 分钟步行）
建造时间：1876—1884 年
建筑师：约瑟夫·齐克特（Josef Zitek）、约瑟夫·舒尔茨（Josef Schulz）

新文艺复兴/ NEO-RENAISSANCE

鲁道夫音乐厅坐落于犹太区边缘，伏尔塔瓦河的河堤旁，与布拉格城堡隔河相望。19世纪末，奥匈帝国皇帝弗朗茨·约瑟夫一世鼓励文化复兴，提倡捷克民族的独立和自主性，使其能媲美巴黎和维也纳。鲁道夫音乐厅的兴建成为捷克物质和文化景观复兴的象征。音乐厅由日耳曼和捷克银行家资助建造，建筑师也是设计国家大剧院的约瑟夫·齐克特和他的学生约瑟夫·舒尔茨。建筑于1884年完工，为新文艺复兴风格。修建时，圣维特大教堂的西侧中殿正在施工，在天际线上还看不到。

鲁道夫音乐厅由两部分组成，南半部是音乐厅，也是建筑的主入口；北半部是画廊和展厅，现在主要展出当代艺术作品，由西面的河堤进入。音乐厅庞大的体量形成一个独立的街区，并采用三段式对称布局。与国家大剧院一样，音乐厅的四个立面各不相同，通过不同的古典主义细节进行表达，比如侧面上的浅双壁柱在正面中央弧形墙上变成几乎高浮雕的圆双壁柱，使中央体量似乎向外爆发，这种表达的转变增加了立面的雕塑感，以及对节奏和虚实比例的感知。音乐厅所蕴含的文化力量通过这种雕塑感的围合成为城市空间的一部分。回忆起庄园剧院的投影，但反过来——它是从内部膨胀的——观众会从流动的楼梯上向四面八方散去。

在庆祝斯拉夫传统的背景下，鲁道夫音乐厅屋顶檐口石栏杆上放置了具有真人比例的雕像，包括捷克和日耳曼艺术家——一群伟大的人物站在屋顶上集体庆祝他们的文明。罗马式偶像的绘画再现、巴洛克人物的神话表达、19世纪晚期工业化帝国人物以先进的人文主义，在人的空间和尺度上代表了那个时代布拉格个人主义和文化成就的高峰。

这座文化综合体内部有众多厅堂，装饰清雅华丽。科林斯柱、大理石饰面、彩绘墙壁和雕塑组合在一起，视觉丰富又充满细节。室内最大的厅现名为德沃夏克厅，是欧洲最古老的音乐厅之一，有1200个座位。1896年1月4日安东尼·德沃夏克在此指挥了捷克交响乐团的首演。

鲁道夫音乐厅在20世纪20—30年代被用作捷克斯洛伐克共和国的众议院。在纳粹占领及共产主义时期，音乐厅恢复了其文化功能。20世纪90年代，音乐厅前的场地被命名为扬·帕拉赫广场，以纪念在1968年因反对苏联入侵而自焚的查理大学哲学系学生。音乐厅现作为捷克交响乐团和鲁道夫美术馆的主场地，每年一度的布拉格之春国际音乐节也在这里举行。

... along the newly built riverfront, a symbol of a society on a representative plinth, standing across from the Prague Castle.

The establishment of the Rudolfinum marked a symbolic renewal of the surrounding physical and cultural landscape. Emperor Franz Joseph I, at the end of the 19th century, encouraged the cultural renewal, the independence and autonomy of the Bohemian people to rival Paris and Vienna. As part of the construction of the embankment, it was to add to the new image of the riverfront.

Funded by managers of the German and Czech Savings bank, it was completed in 1884 by the same architects who worked on the National Theatre. Built outside the core of the Jewish Town, it became integrated into the new city over the following decades. It was placed upon a new, elevated square framed by other civic institutions. At this time, the Western nave of St. Vitus Cathedral was under construction and not yet visible on the skyline. From Jan Palach náměstí, the square in front of Rudolfinum, it is still possible to see the Castle and Cathedral from an angle as it appeared in centuries past, without the Neo-Gothic towers.

The building is composed of two parts which are assembled into a single urban block. The South-facing half, facing the square, is the Concert Hall. The Northern half, with its entry from the western embankment, is the Exhibition Gallery. The single construction was built upon an artificially raised ground, which is now at the level of the city. At the time of its construction, older houses on their lower original grounds still stood around it.

Similar to the National Theatre, the

Rudolfinum is composed of four distinct sides. Each facade is particular. The main entrances face the square and the Castle, and the service entries are on the other sides. The convex wall, the entrance to the concert hall, is framed between rectangular volumes at its sides. The separate forms are articulated through their different Neo-Classicist details which are particularly seen at the edges where the curved wall and the rectangular volumes meet. The double pilasters on the side volumes become high-relief, nearly rounded pilasters, on the central curved wall. This shift in expression heightens the sculptural quality of the wall, as well as the perceived rhythm and ratio between solid and void proportions.

The central volume appears to burst outwards. The cultural power contained within the concert hall becomes part of the city space through this sculpted enclosure. Recalling the projection of the Estates Theatre, but inversely — it is bulging from within — its cultivated audience is to spill out in all directions, down the flowing stairs.

In the context of celebrating Slavic heritage, the Rudolfinum introduces sculpted, human-scale figures of artists — both Czech and German. A colonnade of great men, in collective celebration of their civilisation, stands upon the roof cornice which is also a stone balustrade. From the pictorial representation of icons in the Romanesque, to the mythical representation of figures in the Baroque, the late 19th century figures of the industrialised empire, in a display of advanced Humanism, represent the pinnacle of individualism and cultural achievement — in the space and the scale of the people. Originally, a statue of Antonín Dvořák was to stand on the stairs with the people. It was placed instead at the base of the stairs, looking up.

Within the multifunctional cultural complex are various halls, including a 1200 seat concert hall, a large, two-level foyer facing the square, meeting rooms and galleries under naturally lit, glazed ceilings. There are ornate details, painted walls and chandeliers, Corinthian columns, and marble finishes throughout the interior.

The Rudolfinum was used as the Parliament during the first Czechoslovak Republic in the 1920s and 1930s. It returned to its cultural function, as the seat of the Czech Philharmonic and Rudolfinum Gallery, during the German occupation and the Communist period. In the 1990s, the square was named after philosophy student Jan Palach, a symbol of the resistance to the Soviet invasion of 1968.

布拉格中央火车站
Prague Main Train Station

地址：Wilsonova 300/8, 120 00 Praha 1, Nové Město
交通：Metro C **Stop** Hlavní Nádraží
建造时间：1901—1909 年
扩建时间：1972—1979 年
建筑师：约瑟夫·凡塔（Josef Fanta）
扩建：扬·博灿（Jan Bocan）、阿莱纳·什拉姆科娃（Alena Sramkova）等

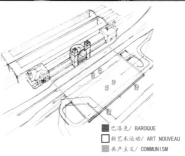

■ 巴洛克/ BAROQUE
□ 新艺术运动/ ART NOUVEAU
■ 共产主义/ COMMUNISM

中央火车站是布拉格最大的火车站。在1871年，它原本计划以新文艺复兴风格——国家大剧院和鲁道夫音乐厅的风格设计建造。但由于建筑实际建造在1901—1909年间，已经到了20世纪，一个新时代的时间点上。因此，为了避免与"旧的"和"既定的"复兴风格相关联，并反映不断变化的时代精神，火车站最终以当时流行的新艺术风格呈现。从某种意义上，这是一个社会夹杂在两种文明之间的表现，是哈布斯堡帝国最后岁月的挽歌，试图满足人民变革的需求，消解那时代出现的紧张的社会局势。

车站位于老城以东，并紧靠城市，面对一个大的花园绿地。车站建筑包括一个中央大厅和两个带有塔楼的长翼，面朝城市并连接着月台。巨大的拱形玻璃钢构入口强调了建筑的宏伟性。建筑首层为候车大厅，上部是管理部门、办公室和服务空间。火车月台上覆盖着大型的双筒钢架和玻璃拱顶，带有弹性销连接的弧形钢墩支撑着钢桁架玻璃屋顶，展示了工业时代先进的材料和建造工艺。

新艺术风格以拉毛粉饰与玻璃、彩色陶瓷和精美金属制品相结合的方式出现在立面上。这些装饰和细节给予这个庞大的新古典主义建筑一个新的形象。中央大厅的立面体现了这些形象和材料表达：雕刻精致的浮雕，仿佛要从正面脱颖而出。车站原始入口大厅的中央穹顶，用金字展示"PRAGA mater urbium"（布拉格，城市之母）。自然主义的花朵和叶子图案以及年轻女性的形象表达了古老的神话和未来。表现罗马和哥特起源的金色插画被重新用于新旧的融合。古今城市符号的结合代表了乌托邦、神灵、怀旧、未来和现代工业社会融成为一体。它将生命视为一个有机体，是城市一种装饰性的、神秘的代表。如果你对新艺术风格着迷，车站中的范托瓦咖啡厅还保持着百年前的原汁原味。

在布拉格地铁和高速公路建设的20世纪六七十年代，中央车站进行了扩建。增建部分是个巨大的玻璃钢构筑物，与老建筑连接在一起。玻璃圆柱体的楼梯以新的形式呼应着新艺术风格时期的塔楼。下层的进站通道顶面被挖空一块，可以透过半圆洞口看到原大厅精美绝伦的新艺术风格的穹顶，历史与当代就这样又相遇。增建部分通过三个主入口将火车站的交通上下连通：从快速路可以通达原新艺术运动风格的老建筑，从地铁可以上到新建的多功能候车大厅，从公园也可直接进入新建大厅。因交通基础设施连接完备，运行良好，中央火车站现今仍是布拉格城市的中心交通枢纽，也是欧洲主要的门户之一。

... upon the demolished city wall, a central connecting point embraces a new cultural promenade.

At the time of its construction, one arrived immediately upon the threshold of the city of Charles IV, at a cultivated garden in the midst of a modernising society. Like the Masaryk Train Station, Prague's Main Train Station doubled as a space of social and cultural life. The main facade of the station, parallel to the promenade, faced into the city.

The Main Train Station is part of the necklace of cultural landmarks which replaced the Baroque reconstructions of the New Town Walls. The city wall was far too massive and fortified to have relevance in a society interconnected by rail. In 1871, the train station was originally built in the Neo-Renaissance style by Vojtěch Ullman, but with its re-construction between 1901 and 1909, and in the spirit of the changing time, the building was redesigned from the established style. By the early 20th century, it was no longer the same society thus to avoid association with the "old" and established Revivalist styles, the Main Train Station was finished in the Art Nouveau by architect Josef Fanta.

As all other important national institutions before, it was placed in a symbolic location. Upon this former rampart, now stretched a long city boulevard lined with cultural spaces, parks, cafés and institutions as well as middle class housing. Facing three directions, the building projected its function as a central point of access between the city and its extended regions.

The station building was set behind a garden park of flowers and a swan lake. The building consists of a 150-metre-long central hall with two wings and towers that face up and down the promenade and into the city. The building reinforces its institutional grandeur with one monumental representative entrance hall for all. Upstairs, there are offices and administrative and service spaces for the management as well as for the workers of the station. The train platforms are covered with two 33.3m wide, 18m tall and 233m long barrel vaults of iron and glass. This construction was engineered and built by a bridge making company. Curved piers with flexible pin connections support the glazed truss roof in a display of the advanced construction methods and materials of the new industrial age.

In the Art Nouveau style, sgraffito combined with glass, coloured ceramics and fine metalworks were brought to the facade. The details and ornamentation gave a new expression to the already defined volume of the massive Neo-Classical composition. The facade of the central hall captures the range of these figural and material expressions, carved from glass and stone with delicate and high relief. Their forms appear as if they are slipping free from their base. To some degree, this was the expression of a society, caught between two civilisations. It was part of a trend in the final years of the Habsburg Empire to satisfy the peoples' demand for representation and change, in an effort to absorb the social tensions of an emerging new era.

Perhaps the most impressive moment of the Art Nouveau style is found within the central domed space of the original entry hall of the station, with its magnificent display of Praga Mater Urbium: "Prague, Mother of Cities". Naturalistic motifs of flowers and leaves and figures of young women express both myths and a possible future. Golden illustrations of Romanesque and Gothic origins are modernised to fuse the old with the new. The combination of ancient and

modern symbols of the city represents utopia, nature, nostalgia, progress, and the modern industrialised society as one. It is an ornamental, mythical representation of the city which sees all of life as one organism.

Severe damage was caused to the original building and its main facade during the metro and highway construction in the 1960s and 1970s. An underground waiting hall was then built as the new entrance point from the park with the city highway placed upon its roof. Now there are three main entries into the train station: from the highway (which cuts across the main facade), from the partially demolished park (now accessed through a massive glass and steel wall of the underground hall), and from the metro (below the highway). It remains the central point of the city, the largest train station and largest Art Nouveau building in the country, and one of its main European gateways. The long-awaited renovation of the building began in 2006 and continues today.

市民会馆
Municipal House

地址：Náměstí Republiky 5, 111 21 Praha 1, Staré Město
交通：Metro B **Stop** Náměstí Republiky（3分钟步行）
建造时间：1905—1911年
建筑师：安东尼·巴尔沙内克（Antonín Balšánek）、
　　　　奥斯瓦尔德·波利夫卡（Osvald Polívka）

■ 哥特/ GOTHIC
■ 新文艺复兴/ NEO-RENAISSANCE
■ 新哥特/ NEO-GOTHIC
□ 新艺术运动/ ART NOUVEAU

市民会馆也许是欧洲新艺术风格最好的范例之一，其建于1905—1911年间，奥匈帝国的鼎盛时期。建筑放弃了早期的新古典主义，意在以一种强有力的综合方式来整合公民社会以及贸易和行会。作为一个公共地标，同时也作为城市进步的代表去体现20世纪初的时代和人民。

市民会馆位于老城区最重要的节点：皇家大道的起始点，火药塔旁边。建筑基地上前皇家法庭大部分建筑在19世纪末已被摧毁。作为一座公民建筑，建筑师安东尼·巴尔沙内克和奥斯瓦尔德·波利夫卡，包括捷克多位著名艺术家和设计师共同参与，展示了当时最先进的技术和工艺，歌颂了集体劳作成就带来的社会凝聚力。市民会馆是那个时代捷克民族复兴的结晶，同时也是对当时迅速发展的捷克政治独立运动的反映。

建筑不规则四边形的布局占据了一个完整的街区。从建筑背后看，有六层高，被街道和商店包围着；但是从其主入口看，市民会馆看起来似乎只有两层高。界面的尺度变化使其与周边的城市肌理融为一体。市民会馆具有纪念性的入口挑台山墙上方为半圆的穹屋顶，穹顶内覆贴着上釉的马赛克。从正前方看，穹顶将马赛克的光影投射到城市当中，仿佛在诉说"向布拉格致敬"。

作为一个完美的建筑艺术品，市民会馆全面展示了那个时代的技艺。在建筑内外可以找到各种最精美的材料和手工艺细节，每一部分都是大师的得意之作。阿尔丰斯·慕夏和马克斯·什瓦宾斯基等名家创作并讲述了斯拉夫民族故事的绘画，模糊了历史和神话之间的界限。

尽管新艺术风格倾向于有机自由的表达，但市民会馆内部的布局却并不灵活。虽然建筑外立面和内部装饰呈现出了一种新的、无与伦比的品质，但中上阶层的社会仪式、历史惯例和公认秩序还是深深扎根在建筑的平面形制当中。场地中不规则形状被内部近乎完美的对称形态消解。建筑为广泛的社交活动提供了包括礼堂、音乐厅、宴会厅、咖啡厅、餐厅在内的多种公共的空间。在那个时代，仍主要服务于社会的中上阶层。建筑主空间是斯美塔那音乐厅，得名于捷克著名民族音乐家贝德里赫·斯美塔那。在1918年10月28日，该音乐厅是捷克斯洛伐克宣布独立的地点。

建筑虽然在布局及结构方面并无太多突破，但是在设备科技方面远远领先于时代：市民会馆配有电梯、集中供暖系统、通风系统、除尘系统、内部通讯系统，还有配备制冰机和冷藏室的厨房。

在今天，市民会馆仍然是布拉格一个充满活力的地标，因其精美的细节、卓越的样式和品质受到广泛称赞。

… at the main gateway into the Old Town, upon Royal foundations, a triptych of cultural heritage.

It is perhaps the best example of the Art Nouveau style in Europe, built at the peak of Imperial-era society. Abandoning the applied Neo-Classicism of earlier periods, the Municipal House aimed to re-integrate civil society, its trades, unions and symbols, in a powerfully synthetic way, representative of the time and the people of the early 20th century.

A modern material palette and crafts are

showcased, in a celebration of social cohesion of the achievements of collective work. This was in part a reaction to the growing development of political independence movements, which have rapidly spread over the past few generations. At this representative and symbolic spot, an attempt was made to re-ground the established authorities into the city through the linking of 600 years of heritage in a single site.

Founded by the Municipal Council as a public house, it was built in the years between 1905 and 1911 as a representation of the city's progressive heritage. The architects Antonín Balšánek and Osvald Polívka were to have also built two more institutions for the state, in the same style, on the Lesser Town riverfront across from the Rudolfinum. These however were not realised due to the First World War.

The Municipal House's monumental entrance balcony, with a curved, arched gable roof, is crowned with a glazed dome. Seen from the front, it projects its mosaic entablature into the city with the depiction of an "Homage to Prague". It is situated at the gates of the Powder Tower, along the Royal way axis, at the most significant intersection into the Old Town. The former Royal palace was destroyed, but its memory was symbolically rebuilt into the new composition. The three elements together form this site-specific triptych of Prague's civic and cultural heritage: the old city wall (with its 15th century Gothic gate and 19th century Neo-Gothic roof), the Municipal House (its 20th century face), and the reconstructed memory of the Renaissance Royal Court (its bridge and passage between).

The building's irregular quadrilateral form is a complete city block. From its main entrance, the Municipal House appears to be only two storeys high, while from the back it is a 6-storey building. Wrapped with streets and shops on the back side it integrates into the surrounding city fabric by changing its scale accordingly.

Despite the organic and curving free expressions of the Art Nouveau style, the internal arrangement is not as flexible. The strangely shaped site is resolved internally in a near perfect symmetry. The rituals of upper middle class society were deeply established within the building plan. Historical conventions and accepted orders were inscribed throughout the carefully arranged sequences and organisations of the rooms. It offers multiple public halls, contains an auditorium, a concert hall, ballrooms, a cafe, restaurants, and a range of other spaces for varying social events — presumably for a wide range of social groups. Yet it accommodates predominantly the higher classes of society.

Nevertheless, as Gesamtkunstwerk, the total display of the artistic culture of society is its primary theme: endless handcrafted details and the finest materials from all the industries are found inside and out. Paintings and murals by Alfons Mucha, tell the story of the nation and people, blurring imagery of history and myth. Every detail, as was intended, is a masterpiece. The facade and the interior finishes represent a new representational and material richness and are of unparalleled quality. The construction methods and engineering used were advanced and innovative, and multiple layers of functions are spatially overlapped on different floors in an inventive way.

Today it is still in use and remains a vibrant landmark with exceptional restaurants, cafés and concerts. Its meticulous and beautiful details are greatly appreciated for their quality and range.

现代主义时期

1918—1938—

　　这个时代也许起始于1899年弗洛伊德《梦的解析》，或者起始于1915年爱因斯坦对时空的再定义，又或者起始于1910年斯特拉文斯基的《火鸟》，还可能起始于1918年马列维奇的《白底上的白色方块》，或可能在1887年马内斯美术家联盟创立之时，就已经开始萌芽。在两次世界大战间隔期间的1918—1938年，也就是捷克斯洛伐克第一共和国时期，民主多元化开启，城市发生变革。在现代民主的人文主义景观中，所有人都致力于构建一个开放的氛围，每个人与他人进行潜在地互动，既保持独立又相互依存。

　　随着第一次世界大战后奥匈帝国的衰亡，新的捷克斯洛伐克第一共和国诞生，教会和王室失去了权力。新的政治秩序和空间结构致力于建立一个新型社会。权力重新面向市场和人民，在集体和共识中，自由个体有了新的社会角色。

　　这个时期经济蓬勃发展，思潮风起云涌，服务于社会各阶层的学校、图书馆、住宅等民用建筑和文化基础设施开始兴建。在这个多语言、多民族的社会中，不断增长的经济推动了富有活力而多元的文化景观建造。

　　当时的捷克斯洛伐克占奥匈帝国21%的领土，却继承了帝国三分之二的工业生产能力，成为欧洲工业化程度最高的国家之一。工厂与集市成为高效生产力的标志。汽车、电影和飞机等新技术的发展，以及日用品的工业化生产促进了人们对抽象艺术的开放态度。这个时代的特质是消除装饰、大规模生产、应用现代新材料和推广标准模数构件。

　　第一次世界大战后，与"现代主义"相关联的风格很大程度上受到建立一个新的、更美好社会的启发。将建筑作为一种手段，通过突出实用性、功能性，运用简单的线条实现美好生活的愿景。在现代主义时期的"风格"谱系中，可以又细分为：装饰艺术派（从新艺术运动象征图像向抽象几何形式转变）、立体主义（一种独特的波希米亚风格，表现了隐含在表面的空间和形式的爆发）、朗多立体主义（从立体表现主义发展的一种短暂捷克风格）、构成主义（对建筑结构的清晰表达）、功能主义（对建筑功能布局的抽象表达），以及现代主义（对建筑结构和功能及相互关系用抽象元素及形式表达）。

　　后来与"现代主义"相关联的是极少主义和简约主义。形式的抽象不是为了

现代主义时期的布拉格
The Modernist City

风格,而是对城市政治史及其既定秩序的激进改造。它利用当时的技术、材料和社会需求,将新古典主义的表达方式转化为新的图像和形式。它强调一种普遍的语言,即抽象的、不对称的、由内而外的立体和平面构成。这也是一种通过生产和培育新的、尚未熟悉的关系和可能性来寻求未来的策略。

在 20 世纪早期,布拉格就是新建筑概念的中心之一。布拉格的功能主义建筑不仅仅是一时的流行,因为最早的功能主义建筑坐落在布拉格新城的历史环境

带形窗
Ribbon Windows

不对称构图
Asymmetrical Compositions

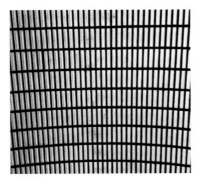

玻璃墙
Glass Walls

无层级结构的场地
Non-Hierarchical Grounds

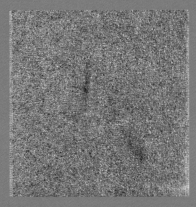

纹理模糊的表皮
Unarticulated Surfaces

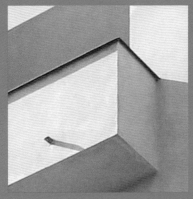

悬挑的体量
Cantilevered Volumes

中,建筑师因此必须遵循传统的历史环境,这就产生了特定类型的功能主义建筑。第二次世界大战与纳粹占领让布拉格的新建筑发展戛然而止,但战后,这种传统又开始继续影响捷克及布拉格的建筑发展至今。

1920 年,捷克斯洛伐克共和国首任总统托马斯·加里格·马萨里克指任斯洛文尼亚人约热·普列赤涅克为布拉格城堡改建项目的建筑师。马萨里克希望这一工程为"国家的典范,一个在君主制中构思和建造出的城堡向民主城堡的转变"。普列赤涅克是奥地利著名建筑师奥托·瓦格纳的学生,马内斯美术家联盟成员,他将 20 世纪的建筑语言与古典的元素结合,使得新的建筑改造与历史环境和谐共生。

普列赤涅克为城堡制定了一整套修缮改建计划,提升了各个空间的实用性和流通性。他对城堡地坪的改造是一个具有现代主义特征的项目。地坪是布拉格城堡中最容易被忽视,且面积最大的人工构造物。普列赤涅克保留并重新调整了地坪图形的几何关系,地坪被当作一个开放的场地被重新塑造:没有权利主导的形象,或者预先设定好的轴线秩序。第三庭院的地面铺装效仿国家剧院的地基选材方法,石料来自全国各地,这个自由的地坪来自其建造的象征意义:以某种开放的表现形式,赞扬新生的多元与统一。布拉格城堡现代主义庭院和花园的重建强调了这一点:平等要素和同一场地中的差异性,是可以互通共融的。

圣维特大教堂旁 18 米高的方尖碑是为了纪念"第一次世界大战"中阵亡的士兵,它以一个现代的、开放的几何形体置于广场中,通过新旧对比,反映出大教堂历史环境的不同片段。这样的建造也体现了在一个中性而复合的空间中,在一个相互分享的地方,人们怎么看待自己。

在现代主义时期的民主治理下,以下这些建筑可以被看作是那个时代最新的表达。

19. 卢塞纳宫
Palace Lucerna

20. 黑色圣母屋
House of the Black Madonna

21. 亚德里亚宫
Adria Palace

22. 展览会宫
Trade Fair Palace

23. 拔佳鞋屋
Bat'a House (Palace) of Shoes

24. 马内斯美术馆
Mánes Union of Fine Arts

25. 芭芭住宅区
Baba Residential Estate

26. 缪勒别墅
Villa Müller

27. 耶稣圣心教堂
Church of the Most Sacred Heart of Our Lord

THE MODERNIST TIME

1918—1938—

   ...*perhaps it began in 1899, with Freud's Interpretation of Dreams, or in 1915 when Einstein redefined space-time. Or perhaps in 1910, after Stravinsky's Firebird, or as Malevich placed his white on white in 1918. Or it could have been in the 1887 Founding of the SVU Mánes Union, simmering beneath the surface for a generation. Certainly between the wars, by 1918–1938, during the First Czechoslovak Republic the beginnings of a democratic plurality began to influence the form of the city. Within the landscape of Modern democratic humanism, all figures are engaged in the construction of an open field of correspondences, wherein each brings their own forces into a potential play with others. No longer structurally separate, each is interdependent within the universally unlimited space and flexibility of the free plan.*

   After the First World War, with the end of the Habsburg Empire there emerged a new political order. An open field, a spatial and structural neutrality, was established for the creation of a new society. The Church and Crown lost power as a new Republic (symbolised by the Prague Castle) reclaimed its place at the heart of the land. Power was reoriented towards the markets and the people. A new social role for the free individual within a collective and common ground emerged.

   President Tomáš Garrigue Masaryk and Castle architect Jože Plečník established — perhaps as a metaphor — the new symbolic ground of the time characteristic of the Modernist project. At Prague Castle the most overlooked construction, and its largest, are the grounds themselves. Without the authority of a dominant figure (Church or Crown), or a pre-determined axis of order, the grounds were re-established as an open field which retained and redirected the collective geometries of all the figures that composed it. This free ground was by virtue of its own construction emblematic — as an interwoven open space — of the celebration of a new plurality in unity. The reconstructions of the Prague Castle courtyards and gardens in the Modern style underscore that the non-hierarchy of elements and differences within a common field is the shared space where individuals unite.

   The obelisk was installed as an open-ended form without closure — an echo of the historical incompletion of the Cathedral (and also a reference to the ongoing and ultimately unrealised Modernist project). Modern time is symbolised as a speculative future (not pre-determined) and Modern Man celebrated (through speculation) as a determining force. This construction (both physical and metaphorical) also represents how people viewed themselves — in a place of mutual sharing — within a neutral and synthetic space.

The booming economy and ideals of the time spread throughout the land, and with them came the construction of civil structures and cultural infrastructure such as schools, libraries, housing and new spaces for all levels of society. In this multi-lingual and multi-ethnic society, the growing economy fuelled a productive and diverse cultural landscape.

Temples of industry and palaces of merchandise, the high cultures of production, crowned in transparency and lightness replaced established and symbolic representations of power. Accelerating technology, such as cars, films and aeroplanes, and the industrialisation of production within every day life enabled a new openness towards abstract art.

Among the "styles" of the Modernist time, we can name: Art Deco (where the symbolic iconography of Art Nouveau is redirected towards geometrical forms), Cubism (a unique Bohemian style expressing spatial and formal eruption implied within the surface), Rondo-Cubism (a short-lived National style redeveloped from the Cubist expressionism), Constructivism (an unadorned expression of a building's construction), Functionalism (an abstracted expression of a building's functional arrangements), and Modernism (an expression, beyond only construction and function, of the interconnected dialogue between abstracted elements and forms).

What became associated with "Modernism" later — as a style of minimalism and reduction — was not its original ambition. The simplification of form was not intended to be stylistic, but rather to be a radical act upon the political history of the city and its established orders. It translated Neo-Classicist expressions into new forms and images using the technologies, materials and social demands of the time. It emphasised a universal language of abstracted and asymmetrical compositions of volumes and planes, inside and out. It was also a strategy for seeking ways to yield the future through the production and the cultivation of new, not yet familiar, relations and possibilities.

The following examples can be considered as a collection of newly emerging expressions, operating in near simultaneity, and made possible during the democratic governance of the Modernist time.

卢塞纳宫
Palace Lucerna

地址: Štěpánská 61, 116 02, Praha 1, Nové Město
交通: Tram 6, 14 **Stop** Václavské náměstí. Metro C **Stop** Muzeum
（5 分钟步行）
建造时间: 1908—1921 年
建筑师: 瓦茨拉夫·哈维尔 (Václav Havel)
工程师: 斯坦尼斯拉夫·贝希涅 (Stanislav Bechyně)

■ 哥特/ GOTHIC
■ 现代主义/ MODERNISM
□ 新艺术运动和装饰艺术/ ART NOUVEAU, ART DECO

卢塞纳宫是新城区瓦茨拉夫广场上一个最古老的，综合商业、娱乐、办公及公寓等功能于一身城市综合体。因沿着主立面拱顶上的弓形桁架形似灯笼并透光，由此取名为卢塞纳(捷克语: 灯笼)。发起、设计并督造这个项目的是瓦茨拉夫·哈维尔总统的祖父地产商老瓦茨拉夫·哈维尔，其与建筑工程师斯坦尼斯拉夫·贝希涅合作一同完成。

建筑始建于 1908 年，但在 1912—1919 年间被搁置，最终在 1921 年完工，这使得修建跨越了奥匈帝国和第一共和国时期。其间，老瓦茨拉夫·哈维尔到访纽约，吸收了美国的装饰艺术派风格元素，建筑设计在保留当时流行的新艺术风格样式和工艺的同时，又加入了巴黎拱廊街的特征，展示出与现代社会文化生活方式相关的理念。这座建筑作为布拉格最早的钢筋混凝土建筑，可以说是法国建筑师奥古斯特·佩雷在 1903 年完成的富兰克林路 25 号公寓的继承者，拥有可以自然采光的玻璃砖顶面和地面，以及布拉格第一座平屋顶露台。

卢塞纳宫将现代建筑综合体融入传统的街景，以一个庞大的、延伸的形态夹杂在相邻建筑之中，并贯通穿过整个街区，室内两层高的内街连接了四条不同的街道。与市民会馆不同的是，卢塞纳宫根据其严格且自我生成的内部组织，适应了周围的街道和场地原有的历史印记，承载并激活了具有活力的城市社会交往。内街穿过整个建筑，在通向建筑内部公共空间的过程中，内部与外部的关系变得模糊。通过提供开放的空间和社交场所，卢塞纳宫达到了内部空间的与城市生活的互动。

在当时，该建筑采用了新的内部布局和立面材料。在吸收新兴风格的同时，它仍然融合了当时的传统风格，比如上层暴露的混凝土外墙和金属板与底层的铁艺和新艺术主题相结合，建筑的立面构图模仿了很多周边传统街坊已有的模式。通过不断使用新的技术和材料，而不是追求隐喻、符号或者象征，卢塞纳宫的混合形式在历史发展中持续下去，至今仍保持着流行的"日常"社会生活中心的地位。

这是一个为中产阶级市民服务的建筑，其目的是营造一个市民社交活动的融合器，一个对外开放的公共空间，因此它不需要明确的象征性外观，内部空间的品质和市民活动功能的满足优先被考虑，比如一半的建筑底层沿街界面都开放给了文化活动或者商铺。剧院、电影院、咖啡厅、商店和大舞厅被几条内街和玻璃顶覆盖的内厅所连接起来。底层咖啡厅采用玻璃墙面将空间对外展示，强化了内部空间与街道的互动。这表达了一种预设的社

会愿景：运用新型的材料和新兴的社会活动创造了一种混合与反射、人与人互动的通透轻盈空间。

尽管经历了百年风雨，建筑内部仍保留了原有的氛围。著名的咖啡厅和电影院在 20 世纪 20 年代早期社会的奢侈气氛下依旧保存完好。建筑修缮也一直在分阶段进行，大理石、黄铜、镜子和玻璃等建筑装饰特征要素大部分还是原貌，一些破损在 20 世纪 60 年代也得到修缮。今天，在主通道中，悬挂着由捷克当代艺术家大卫·切尔尼在 1999 年创作的著名雕像——捷克的圣人——瓦茨拉夫骑着翻转过来的马。2019 年，建筑的屋顶变成了可以举办活动的花园露台，并可以眺望美丽的布拉格城市天际线。

…at the central point of Charles IV's New Town plan, an interwoven passage of modern culture.

Palace Lucerna, being neither conventional nor radical, survived the fall of the Empire at the end of the First War. The city's first mixed use development, a complex of culture, offices and apartments, lasted through one regime into another. Its primary client, the Bourgeois society, remained a key pillar of the Modern era. While retaining the style and craft of Art Nouveau (as well as that of the established Neo-Classicisms), it began to display ideas associated with Modernity.

Closely linked to a social vision which endured beyond the war, the Palace Lucerna was realised in 1921, 12 years after construction began. Located near Wenceslas Square, between Vodičková and Štěpánská Streets, it was initiated and built by Václav Havel, grandfather of President Václav Havel with engineer Stanislav Bechyně. Construction began in 1908, and was put on hold between 1912 and 1919. With Parisian qualities that brought Modern culture and decorative arts into architecture and society, it has carried its core, or its civic purpose, through to the present.

Palace Lucerna integrated a modern complex of many functions within a traditional streetscape. Its long, linear form is squeezed between its neighbours and stretches through a city block. It is composed of two primary volumes made of smaller volumes. The first consists of public spaces on, above and below street level, including 4 underground floors. The second consists of 6 stories of offices and apartments overhead. At the centre of the upper volume is a large courtyard volume with an open flat roof garden and multiple terraces and a glazed roof that brings light into the passages below. Along the main facades, the roof is topped with a bow truss which recalled the form of a lantern and gave the Lucerna its name.

The building was the first in Prague to be made in reinforced concrete. In addition to its new construction technology, it has a novel internal layout and reinvents material expression on its facade. Osvald Polívka, architect of the Municipal House, was also the architect of the facade on Vodičková Street. The upper floors have exposed cast-concrete facades with metal panel detailing combined with the ironwork and Art Nouveau themes below. While it internalised the emerging

trends of building with concrete, it still blended and incorporated the traditional styles of the time. The facade composition reflected the established patterns seen in its more traditional neighbours. By celebrating the materials of new technologies, without added metaphors, symbols or iconoclasm, the hybrid form of the Palace Lucerna allowed it perhaps to endure through a transformative history, and to this day remain a popular "everyday" social centre.

It was meant for the average middle class. Aiming to be a fusion of civil society and more accessible to the public, it did not need an explicitly symbolic facade. Priority was placed on the quality of the interiors and the civic functions offered; with half the total building dedicated to cultural activities. The halls, cinema, cafés, shops, and large ballroom for 4000 people are accessed through an open passage and glass-covered forum. The double-height network of passages at ground level has been expanded through adjacent buildings and now connects into four different streets.

Palace Lucerna adapted to the streets surrounding it and also to the historical imprints beneath it. Inside and outside are blurred as the streets continue through the building for social gatherings. As such, it houses and it activates a dynamic civil society. Within the passage, the café is open to the public space through its internal glass facade. Its interior engages city life, providing communication and activities on an expanded ground level. It is a model of the society envisioned from its start: one of new materials and activities which create a space of mixtures and reflections, transparency, lightness, and interaction amongst people.

Despite decades of disrepair and conflict, its interior atmosphere has been preserved. The renovation and renewal continue in phases. Its celebrated cafe and cinema, luxuries of the early 1920s society, are still in operation. Details in marble, brass, mirror and glass are mostly original with some renovations from the 1960s. Today in the main passage, one can also see the suspended statue installed by Czech artist David Černý in the 1990s of Prince Wenceslas riding upright his upside down horse. The rooftop was made into an accessible garden terrace in 2019 with beautiful views of Prague.

黑色圣母屋
House of the Black Madonna

地址：Ovocný trh 19, 110 00 Praha1, Staré Město
交通：Metro B **Stop** Náměstí Republiky（5 分钟步行）
　　　Tram 2, 6, 8, 15, 26, 41, 91, 92, 94, 96 **Stop** Náměstí Republiky
　　　（5 分钟步行）
建造时间：1911—1912 年
建筑师：约瑟夫·戈恰尔 （Josef Gočár）

☐ 立体主义 / CUBISM

捷克立体主义建筑在现代建筑史上具有独特的影响力，黑色圣母屋则是布拉格最重要的立体主义建筑之一。它位于老城一个三角形街区的顶端，采莱特纳街和欧沃茨尼市场的交会处，城邦剧院与市民会馆之间，矗立在哥特、巴洛克以及新古典主义的历史环境中，并叠合在布拉格的城市肌理之中。

黑色圣母屋的建筑师为约瑟夫·戈恰尔，他是马内斯美术家联盟的成员之一。马内斯美术家联盟是19世纪末成立，由布拉格先锋美术家和建筑师组成的艺术同盟，试图推广适合于新时代的表达形式，而不是参考历史上的古典复兴风格。黑色圣母屋与其对面的市民会馆是布拉格历史上首次在同一时间和地段上建造的两种完全不同风格的建筑。与新艺术风格的市民会馆不同，黑色圣母屋是一个四层高并带有双层屋顶的现代建筑，其屋顶的层叠形式呼应了邻近巴洛克建筑的屋顶，而精致富有雕刻感的立面呼应了基地的形状和附近庄园剧院平坦卷曲的立面特征。

黑色圣母屋是空间和时间的纽带：无论是在既有体制之内，还是在历史前进的突破点上。它强调了新兴的现代主义运动在绘画上与三角形和多面图形的关联。作为一种新精神的代表，它站在了既有建筑风格的对立面，鼓励将新兴风格融入哥特和巴洛克的环境之中，以此证明将新形

式整合到历史语境中的可能性。

　　黑色圣母屋由于受到场地历史遗存的影响，呈现出住宅建筑的特征。它在街角采用了一种内敛的表达形式——既不突兀也不平淡。建筑采用了既有的砖石和铁梁的建造技术，但其内部的布局是非传统的，并且其立面展示出了一种新的风格表达。

　　建筑一个独特之处在于窗户的细节和其立体表现力。这种立体形式的窗户由三个面组成，这种多面的体量反转了传统的形式关系，呈现出从立面中迸发出来的感觉，蕴含了这一历史时刻的本质——新旧事物之间的紧张关系。

　　近距离观察正立面，壁柱不再遵循历史的范式；相反，它突破了既定的规范，以立体主义的几何形展开特有的表达。建筑的外观一方面融入其所在的语境当中，另一方面通过其所有的要素映射出一种从内而外的张力。值得关注要素还有建筑东北角上的雕塑：黑色圣母。这个来自被拆毁的巴洛克房屋的细部被保存并整合到新的建筑立面之中。

　　现在，建筑下层设有立体主义主题的商店、博物馆和咖啡馆，上层是艺术家工作室和办公室。每个楼层都非常开阔：内部的空间不会被柱子干扰。2005年，这个可能是世界上独一无二的立体主义风格咖啡厅重建时，由于原有设计图纸无一幸存，因此采用了1912年以来咖啡厅的内部陈设的黑白照片来作为参考，室内原初那些细节，比如楼梯、家具、灯具等都以原有立体主义的形式和风格被复原。

… at the tip of a triangular block, a fold within the city fabric.

The House of the Black Madonna is a doubled hinge in space and time: in the road between established institutions, and at a shifting point of history. It emphasises the triangulated and faceted image associated with the emerging movement of Cubist painting. As a counterpoint to the established architectural style, it is representative of a new spirit. It blends an emerging style into the Gothic and Baroque environment, and demonstrates the possibility of integrating new forms into historical contexts.

There are other Cubist buildings in Prague and throughout the country, but this is the most significant within the Old Town. It was built between 1911 and 1912 by architect Josef Gočár, who was among the members of the SVU Mánes Union. This group of artists and architects, sought to promote new forms of representation more appropriate to the Modern age than the historicist and Neo-Classicist references of the Revivalist styles.

This was the first moment in the history of Prague where two entirely different styles were built at the same time and place. The House of the Black Madonna, seen opposite the Municipal House, was completed in the same year. The House of the Black Madonna is informed by the heritage and history of the site and blends with its surrounding buildings. It introduces into the corner block a restrained form of expression which is neither disruptive nor familiar. At the intersection of Celetná Street and Ovocný Trh, it is a single volume with 4 floors and a two level roof. The roof was sculpturally layered to mimic the neighbouring Baroque roofs, and the delicate fold of the facade echoes both the shape of the site and the flattened curved facade of the neighbouring Estates Theatre.

Its innovative facade exhibited a new expressive style. The window details and their three-dimensional appearance are a unique feature of the building. The layered and faceted massing of these windows inverts conventional relationships of window to wall. They appear to be bursting out from within the facade surface. Despite the massive masonry of the wall, its openings appear more solid than the stone. The three sides of the building are wrapped in a single bent facade in an echo of these triangulated forms. At a smaller scale, the pilasters no longer follow historical motifs. Rather, they exhibit a characteristically Cubist expression of unfolding geometries and again echo the sculptural, disruptive play against established norms. The appearance of the building, on the one hand blends into its context and on the other, it projects through all of its elements, a force of rupture from within. This captures the essence of this moment in history: a restrained but emerging eruption of the new from within the old.

Its interior layout was also unconventional. The open plan of each floor provided a unique feature: an interior space uninterrupted by columns. The combination of a wall-free space and the projecting volumes of window, creates an exceptional expanse inside the first floor café. The House also contains a Cubist design shop on the ground level and on the upper floors, artist studios and offices. In the 2000s when the café was reconstructed, it was not known which were the original colours, but the original details, its Cubist staircase and furniture were restored in their style.

An element to note is the sculptural detail on the Northeast corner: The Black Madonna. This remnant of the demolished Baroque house was preserved and reintegrated into the new facade, and gives the house its name.

亚德里亚宫
Adria Palace

地址：Jungmannova 31, 110 00 Praha 1, Nové Město
交通：Metro A, B **Stop** Můstek（3 分钟步行）
建造时间：1923—1925 年
建筑师：帕维尔·亚纳克（Pavel Janák）、约瑟夫·扎克（Josef Zasche）

□ 朗多立体主义／ RONDO CUBISM

亚德里亚宫也许是朗多立体主义（Rondo：表面采用光滑弧形元素）建筑中最引人注目的一座。它将意大利文艺复兴和未来主义的意象与波希米亚现代主义融为一体，产生了前无先例的形式，可以被看作第一共和国波希米亚精神的代表。这种形式不是为了怀旧，而是从根本上重新定义历史上继承的既有范式，并代表了对新的合成事物的惊人知觉，将遗产和未来融入一种新的和谐形式中。

亚德里亚宫建于1923—1925年之间，朗多立体主义风格的鼎盛时期，由建筑师约瑟夫·扎克和帕维尔·亚纳克（立面设计）为意大利亚得里亚保险公司设计建造。建筑地上八层，地下三层，钢筋混凝土框架和砖砌体抹灰结构。尽管其外观看起来是一个坚实的砖石体量，但框架支撑的内部结构提供了新型的开放式空间，并继承了卢塞纳宫中带有通廊的多功能街区建筑类型。建筑沿着道路边线有一定的退界，这不仅是为了留出广场的公共空间，同时也展现了建筑视觉冲击力以及屋顶上跳跃的雕塑形象。

与安娜王后的夏宫坚实的石头基座做法相反，亚德里亚宫的底层采用轻盈的玻璃，作为城市生活的展示空间，柱间距较宽，给人通透的感觉。这与建筑上层的窄窗间距形成了体量上的鲜明对比。外立面满布密集的圆形和正方形的装饰图案嵌入在建筑表面，看起来像是通过螺栓和紧固件将装饰面板固定到框架上一样，使厚重体量的建筑得到了一种精神性的升华。建筑在正立面出挑一个巨大的二层露台，面对着开放的广场。建筑首层内部形成了一条两层高的通廊街和一个巨大圆厅，并设有商店和咖啡厅。扬·什图尔萨和奥托·古特弗罗园德等人为它设计了华丽的室内装饰：奥地利棕红色大理石饰面、手工打磨的黄铜窗框、马赛克拼花地面、枝形吊灯、环形的灯泡营造出了那个年代独有的奢华感。

因其强调掩盖结构的表皮，对历史元素富有创意地抽象和使用，亚德里亚宫也许看起来像今天的后现代主义。尽管建筑采用了文艺复兴、巴洛克的主题和历史意象，但亚德里亚宫并没有回归历史主义。也许这就是为什么它的外表如此地困扰勒·柯布西耶。

1958年，莱特尔纳魔力（第一个多媒体表演剧院）在该建筑的地下剧场里成立。1989年之前，这里是瓦茨拉夫·哈维尔总统领导的公民论坛总部。如今，亚德里亚宫是布拉格电气公司的办公中心。

…along the line of cultural institutions around the former Old Town Wall, a modern stitch displays its woven fabric.

Its synthesis of Renaissance, Baroque and Modern ideas yields a form which today has no precedent or antecedent. The Adria Palace can be seen to represent the international Bohemian spirit of the First Republic. It fused imagery of the Italian Renaissance and Futurism with Bohemian Modernism, in this, perhaps the most striking of all Rondo-Cubist buildings. The unconventional use of historical imagery was not in the interest of nostalgia or anarchy, but to present a startling perception of a new synthesis. It was representative of the enthusiasm of the newly emerging possibilities of the time, and the unexpected combinations that may result. These new interpretations of history and Modernity put heritage and the future into a new form of harmony.

At the height of the Rondo-Cubist style (which lasted only briefly), the Adria Palace was built for an Italian company, Riunione Adriatica di Sicurtà, between 1923 and 1925 by architects Josef Zasche and Pavel Janák who designed the facade. The typology of a multi-function block building with passages was inherited from the Palace Lucerna model. Unlike its predecessor's relatively neutral ornamental facade, the Rondo-Cubist palace was built to express the spirit of novelty in the autonomous Republic. This avant-garde image was also to represent a society of international influence, which combined the values of culture, history, exchange and invention with a politics and enthusiasm for symbols of a new type.

Along the former Old Town wall, at an important intersection within the city's Gothic structure, the 8 level building occupies the mass of a city block. It expresses an institutional monumentality on Národní Street, and a subdivided scale of four towers to match the facing buildings on Jungmannová Street. The building was set back from the street line not only to give pronounced form to the public space of its forecourt, but also to emphasise its explosive character and its top-heavy expression.

The lower level of Adria Palace, in an inversion of the stone plinth from Queen Anne's Summer Palace, was a lightweight space of glass which became a public gallery of city life. The large openings between columns on ground level gave the impression of an airiness in high contrast with the upper volume. The mass of the heavy block and towers was sublimated and subsumed in the massive and dense ornamentation of the facade. Patterns of circles and squares embedded in the stone skin appear like bolts and fasteners attaching panels to the frame. There is nearly no space left but ornament.

On the line between the two volumes is a perceived tectonic schism. The bulky package overhead appears to slide, as if gliding on the rails of a two-level plinth of glass, and echoes the expressions of Italian Futurism and industrial design. Their displacement produces a large terrace on the second level, facing the open forecourt, and a split level form of private-public space. Within a two level passage space, which includes shops and cafés and entries to the theatre below and offices above, there are hand-built brass framed windows with large panes of glass, marble finishes and recessed lighting.

It may appear today "postmodern", by emphasising its surface as a mask over its structure and for its whimsical reuse and abstraction of historical elements and references. But, Adria Palace does not fall back to historicism, however, despite using Renaissance motifs and referring to historical imagery. Perhaps this is why its appearance confused Le Corbusier so much.

In 1958, Laterna Magika (the first multimedia performance theatre) was established in the underground theatre space. Today the building also serves as the central office of the Prague Electrical Company.

展览会宫
Trade Fair Palace

地址：Dukelských hrdinů 47, 170 00 Praha 7, Holešovice
交通：Tram 1, 2, 5, 6, 14, 17, 25, 41, 93, 94 **Stop** Veletržní palác
　　　（1 分钟步行）
建造时间：1925—1928 年
重建时间：1986—1995 年
建筑师：奥尔德日赫·季尔（Oldřich Tyl）、约瑟夫·富赫斯
　　　（Josef Fuchs）
重建：SIAL 建筑公司（SIAL Group of Architects）

☐ 结构主义 / CONSTRUCTIVISM
■ 共产主义 / COMMUNISM

展览会宫位于布拉格老城以北的工业区：霍拉舍维采，由建筑师奥尔德日赫·季尔和约瑟夫·富赫斯共同设计完成，为钢筋混凝土结构。建筑始建于1925年，于1928年10月为庆祝捷克斯洛伐克建国10周年而落成揭幕。展览会宫是布拉格最早的，也是欧洲最早的功能主义建筑之一。展览会宫将工业（建筑）的尺度带入了城市居住区，建筑庞大的体量占据了整个街区，成为了当时世界上最大的功能主义建筑之一。当勒·柯布西耶第一次看到这个建筑时，也惊讶其巨大的形态和尺度，并揶揄说："这是一座有趣的房子，但不是一座建筑。"

作为最初定位为产品交易的建筑，无论形式还是功能上，展览会宫都反映了那个时代：不需要历史符号或象征主义，通过现代技术来表达理性、抽象、效率的时代特征。新的材料和结构形式作为一种风格，以过去截然不同的形式与周围的历史街区产生对话。

与周围注重装饰的历史建筑相比，展览会宫朴素而低调的外观显然没有历史参照，唯一例外的是底层商店使用了可活动的织物遮阳棚（现在已拆除）保持了街道的传统生活和节奏。功能决定形式的建造逻辑阐释了功能主义的诉求：简单的几何体块，带形的长窗，由于楼板由悬臂梁支撑，因此无需设立柱子打断连续立面。建筑中有两个巨大的中厅，被设计成传统与创新、水平与垂直相互对比的两个空间。第一个横向布置的大厅是庭院式，建筑四周围合，人并不能进入。另一个垂直布置的大厅是教堂式的，是覆盖玻璃顶的垂直空间，四周悬挑的阳台包裹着大厅，将垂直空间分布和人的活动对外展示。建筑入口空间是一个多层次的中庭，将城市的公共空间延伸到室内大厅。

1941年，展览会宫作为贸易展览馆使用还没到15年，就被纳粹征用作为收罗财产和运送布拉格犹太人出城的据点。20世纪50年代初，建筑被改造为办公空间。1973年，展览会宫被一场大火彻底烧毁，在1986—1995年间进行了重建。如今，它是国家美术馆的一部分，主要展出19—20世纪的艺术作品，其收藏的梵高、毕加索、席勒、克里姆特、穆夏等大师之作及捷克现代艺术作品一定让你不虚此行。

... in a continuation of the cultural promenade, the space of the city opens into a hall of industrial culture.

It was a new type of extended market turned into a city block. Tasked with bringing the industries and trades into the residential fabric of the city, it was not intended to be a luxurious place of contemplation as much as a vast space full of people and products. In both its form and use, it was reflective of its time: abstracting and re-appropriating heritage into an innovative civic purpose.

Placed along the modern-day residential boulevard, at the industrial edge of the city, it extended the historical centre northwards towards Stromovka park and the new Exhibition Grounds. The modern civic structure occupies an entire city block and brings an industrial scale into the residential fabric. Symbolic references to historical architecture, and Roman symbolism were not desired. A celebration of modern technology, rationality, abstraction, efficiency and lightness were sought.

With an emphasis on the possibilities of construction, architects Oldřich Tyl and Josef Fuchs realised the Trade Fair Palace from 1925 to 1928 in reinforced concrete. A relatively unexplored material at that time was used in this building in a creative way. The possibility to cantilever volumes above the ground, and to "dissolve" the appearance of weight was used as a means of stylistic expression as well as a means of dialogue with the surrounding blocks of the city.

Four walls of the city block are made by four vertical slabs which wrap galleries and other functions within. The large building is a landmark which stands as if in its own square. The long plaza in front of the entrance extends the public space of the street into the interior. The palace is no longer a separated "institution enclosed by a wall and private garden", but instead becomes a public opening in the city space. The Trade Fair Palace employs a "Miesian" spatial trick: it sets back from the street in order to gain height. In so doing, it establishes itself as an exception to be seen from all the edges of the lower peninsula.

Opposite the embellished blocks of houses, the Trade Fair Palace was stark and silent in appearance. It expresses Modernity itself without any historical reference. One exception to this was the use of ground level shops with (now removed) operable fabric awnings which maintained the traditional life and rhythm of the street. Otherwise its form was determined by the material construction with a concerted minimisation of ornament. Not only did this have an underlying economic logic, it enabled for continuous ribbon windows across the facade. and a nearly fully glazed wall. As the floors were supported by cantilevering beams, there was no need to interrupt the facade with columns.

The spaces of the two interior halls, and their contrasting morphologies, were put into opposition. They were to be the space of interaction between tradition and innovation; between the horizontal and the vertical. The first hall, a courtyard-type, had an axial layout of a traditional basilica, while the second hall, a cathedral-type, had the scale of verticality of a modern tower capped with a glass ceiling. Space and also people are put on display along the cantilevered balconies which wrap the interior of the vertical hall. The dialogue and difference between these two types of galleries are joined in the foyer. The entrance space is a multi-level atrium which extended the public space of the city into the halls within.

Its intended use as an exhibition hall for the trades did not last even 15 years. In 1941, under occupied Czechoslovakia, the halls were used as a place to collect the properties of Prague's Jews and to transport them out of the city. In 1973 a large fire damaged the building and it was later renovated between 1986 and 1995 (including the parking ramp inserted into the front plaza). Today it is used as a part of the National Gallery of Prague.

拔佳鞋屋
Baťa House (Palace) of Shoes

地址：Václavské náměstí 6, 110 00, Praha 1, Můstek
交通：Metro A, B **Stop** Můstek（1 分钟步行）
建造时间：1927—1929 年
建筑师：卢德维克·科特拉（Ludvík Kysela）、英德日赫·斯沃博达
　　　　（Jindřich Svoboda）

■ 巴洛克 / BAROQUE
□ 功能主义 / FUNCTIONALISM

拔佳是原捷克本土著名的制鞋企业，也是世界最大的鞋业公司（1948年后总部迁到加拿大）。位于瓦茨拉夫广场的拔佳鞋屋在1927—1929年间设计建造，在当时是一座为城市新兴中产阶级服务的鞋类专卖店。这个时期，拔佳公司在捷克斯洛伐克及欧洲其他地方建造了很多这类功能主义的建筑。这种按照捷克兹林拔佳工厂的功能主义原则建造的建筑，在当时被认为是社会进步、经济繁荣、创新精神和社会多样性的象征。

拔佳鞋屋处在布拉格新城区哥特城市肌理中一个不寻常的位置：两个独立广场的交会处，并且是它们之间的通道。建筑是一个地上8层和地下2层的独立体量，并通过电梯使得各层楼面在一个共同的立体空间内实现价值的平等。钢筋混凝土框架的开放结构也使得新古典主义的内部等级和规则被消除。建筑首层被设计为一个开放空间，其玻璃界面作为城市的连接点向两个广场开放，使得建筑融入周围细腻的城市肌理。首层通过地面斜坡衔接了两个广场的不同高差。在高低地面交接处，楼梯（现在是自动扶梯）就像鞋带一样，将不同的水平面缝合在一起。

拔佳鞋屋可以说是拔佳公司强大商业机器的延伸。建筑的钢筋混凝土圆柱浇铸在金属板上，开放式布局灵活性和适应性强，可以容纳任何展品。在设计概念上，每层楼都可以独立变成一间房间、一座商店、一间办公室或一间大厅。若空间完全被鞋子陈列填满，整个建筑物就变成了货物仓库。与展览会宫不同，拔佳鞋屋采用了全玻璃幕墙覆盖整个立面。楼板位置用水平的白色哑光玻璃带区分且遮挡。建筑立面的古典性被非物质化，因为窗户不再是墙上的"开口"，其本身就是墙体。正立面三层及以上整体挑出的墙面又强化了建筑的立体感。

在夜晚，通透明亮的灯光从玻璃幕墙里倾泻出来，成为拔佳鞋屋最好的商业灯光广告。屋顶上醒目的红色"拔佳"标志让人想起曾经在这个地方的花园中生长着玫瑰。

...in the space between two squares, a Gothic passage through the city becomes a Modern service for feet.

It was a building for the middle class and also an icon of a new society. In itself it was a testament to the thriving economy of the time, the spirit of innovation and variety, and the celebration of a culture of manufacturing. Unlike the department stores of many shops, the Baťa Palace of Shoes was specifically for one merchant. In this case, the private markets were internalised within the building, as a Palace for the exchange of merchandise directly between manufacturers and consumers.

Built for the Baťa shoe company, by architect Ludvík Kysela between 1927 and 1929, it was a sales department store to serve the city of Prague. Many such houses were built around the country (and Europe) for the Baťa company. The standardised construction principles from their factory city in Zlín were re-applied in this Palace of Shoes using a repeatable system of technical standards and clear functional hierarchies.

The site is an unusual moment within the Gothic fabric of the New Town. It is at the intersection point of two separate squares, and served as a passage between them. The glass walls at ground level open to both of the squares and the interior becomes a connecting point within the city. The Baťa House fits into the slivers of its nuanced ground (as a good shoe store should). In the 14th century, homesteads with rose gardens laid claims to added land set into Wenceslas square. For this reason the footprint of the property was set slightly forward compared to its neighbour. To emphasise this forward projection, the facade was set back from the street only at ground level.

The neutral, structural frame of this standardised system remained a necessity, but internal hierarchies and rules of the Neo-Classical orders were otherwise displaced. The building was conceived as a single volume, with 8 stories and two basements. The ambition was to make equality across levels with no significant difference in vertical arrangement. A Modern market, with elevators, could be a civic tower. The ground level was conceived as an open space, which aligns with the different ground levels of the two squares. The stairs (now escalators), stitch the levels together, like shoelaces, at this fold in the floor.

The entire facade was covered in a fully glazed curtain wall. Unlike in the Trade Fair Palace, the classical massiveness of the solid facade dematerialises here. The windows are no longer "openings" in a wall, but become the wall itself. The traditional composition of window openings within a solid mass, however, is maintained in the horizontal rhythm of the facade, as strips of white matte glass mark the locations of floor slabs. This also aligns the new form with its adjacent buildings. And significantly, the exposed corner which projects out above the street, accentuates the building's identity as an independent volume.

Hierarchy was effectively eliminated within the interior, both vertically and horizontally. The open plan could accommodate any layout. Its rational grid of flexible, adaptable and neutral internal organisation was composed of circular reinforced concrete columns cast in metal sheeting and an

unimposing grid of square ceiling panels (since covered over). The exposed vertical elements of circulation on the "back" wall of the space (actually its side) allowed for each floor, at least conceptually, to become a room, a shop, an office, or a hall. The sales shop was conceived as an extension of the vast machinery of the business. The variability of content and the durability of design was reflected in both the architecture and the product. The entire space could conceivably be completely filled with shoes, and the entire building converted into a warehouse. The building once included rapid shoe repair, consultations and foot care, offices and a snack bar on the top floor.

Today, it is still an operating business, and the iconic red Baťa sign on its roof recalls the roses which once were growing in the gardens in this spot.

马内斯美术馆
Mánes Union of Fine Arts

地址：Masarykovo nábřeží 250/1, 110 00 Praha 1, Nové Město
交通：Tram 14, 17 **Stop** Národní divadlo（3 分钟步行）
建造时间：1928—1930 年
建筑师：奥塔卡尔·诺沃提尼 (Otakar Novotny)

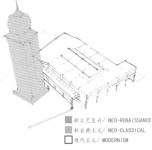

■ 新文艺复兴 / NEO-RENAISSANCE
■ 新古典主义 / NEO-CLASSICAL
☐ 现代主义 / MODERNISM

马内斯美术馆横跨在伏尔塔瓦河上，连接着新城和斯拉夫岛。现代极简的白色剪影与其背后新文艺复兴风格的水塔共同形成一个更大的构图，并与周围古典的历史环境形成鲜明对比。1928年，马内斯美术家联盟的主席，奥塔卡尔·诺沃提尼打破传统，完成了这幢几何体块、带形天窗、开放式的钢筋混凝土建筑。开放的艺术空间被置于城市公共话语中，承载着共同的历史和文化价值。通过对历史的批判，马内斯美术家联盟将自己的政治理想通过建筑转化为城市的结构和语言。

马内斯美术馆的建筑轮廓首先对城市肌理做出呼应。建筑的白色、水平、间断的线性体量和其临近石塔的深色、垂直、连续体量之间形成对比，在"旧"和"新"的对比中形成了一个更复杂的整体。从布拉格城堡眺望，建筑部分隐藏在树木中，并编织在水塔的"历史性"结构中，使古老肌理的城市中有一个模糊的白色图影。同时，这座建筑可以被当成一座桥梁，漂浮在河上，成为一个超现实的景象，这也是对20年前建成的新艺术风格的市民会馆的一种回应。

从立面构成来看，纯白的建筑没有多余的装饰，建筑的窗户被放置在一个排列组合中，如同一幅平面构成。建筑的每个立面都不同，既与周围环境相通，又反映了其内部布局的特征。此时的设计哲学不是立体主义，而是那个时代的先锋语言：抽象和拼贴。建筑内设画廊以及其他制作、讨论和展示艺术的空间。

这幢建筑在1948年后被关闭，上层的室内空间不幸被破坏。1990年，马内斯美术家联盟总部迁至钻石大厦，一个新城的立体主义建筑内。如今，该幢建筑再次成为公共艺术中心，展示来自不同个人和组织的艺术作品，并附设有餐厅、商店、咖啡厅等。

…in a provocative situation — within but against history — a bridge between the city and an island.

Not only was it radically new, but it radically combined the new with the existing. In an ambiguity by design, it provoked the imagination. Multiple front facades produced an inability to place a dominant. The possibility of a plurality of orientations — all framed within one image, one white box, one canvas — was a mirror of the spirit of the time. Through their own long history of critique of the established orders, the SVU Mánes Union institutionalised their own politics into the structure and space of the city.

It is located along the embankment of the river, mid-way between Prague Castle and Vyšehrad. In a break with established tradition,

in 1930, Otakar Novotný completed an open structure of reinforced concrete with skylights and windows of varying sizes and a wall of glass at the main entrance. Conventions were challenged by the stark opposition between the white, punctuated linear volume and the massive vertical stone tower at its side. An open space of art was placed into the public landscape.

Compared with the nearby National Theatre, completed in 1883, it was a strong contrast to the institutions recently planned for the riverfront. Built upon a former mill, it made a larger composition with the Renaissance water tower next to it. Its depth was equivalent to the height of the tower and its framed entry corridor, a colonnade of bent beams, echoed the tower's rhythm. The building responds to the city fabric at all its edges, especially along its street-front on the embankment. Standing over the water, upon the former foundations of a mill, its floor aligns with the ground of the city. The intersection of the Masarykovo embankment and Myslíkova Street was pulled into the gallery through its operable front wall of glass. At this "white box" temple, there was no plinth. In a twist of the Modern time, the land was woven into the temple directly and the palace was flattened level with the ground.

A more complex whole was synthesised through the apparent collision of past and future. Seen from the Prague Castle, partially hidden by trees, and built into the "historical" structure of the water tower, it appears as an ambiguous moment of white within the historically colourful city. The building, as a bridge, also floats as a surreal vision upon the river. It is seen as both a rupture and a possibility. Echoing the Municipal House, built only 20 years earlier, the space between old and new which connects both sides of history is the space of the open passage they share. The space between the "historical" tower and the "new" gallery is a volume, a void and a bridge in one; a framed yet open space; a symbol of the present.

Windows were placed into a composition as they would be placed into a modern canvas. After all, its purpose was to display field-breaking forms. However, the philosophy and its expression at this point developed less from Cubism as from abstraction and collage. It was not necessary to decorate the facade with ornament, but to leave "white space" open for interpretation. Each facade was specific and communicated with its different surroundings and reflected the character of its internal arrangement. Within the building were galleries and offices, terraces and a restaurant, and other spaces for the making, discussing and display of art.

The building was shut down during the Communist period and unfortunately its original upper floor interiors were destroyed. In 1990, the SVU Mánes Union headquarters moved to the Diamond House, a Cubist building in the New Town. Today the Mánes Union of Fine Arts building is again a gallery space that displays works from various individuals and schools of art.

芭芭住宅区
Baba Residential Estate

地址：Na Babě 1801/8, 160 00 Praha 6, Dejvice
交通：Bus 131 Stop U Matěje（7分钟步行）
建造时间：1929—1932年
规划师：帕维尔·亚纳克（Pavel Janák）
建筑师：捷克斯洛伐克制造联盟（Czechoslovak Werkbund）

☐ 现代主义/ MODERNISM

在 1932 年的欧洲住宅展览会上，位于布拉格历史城区西北面的芭芭住宅区引起了人们极大的关注，最终被写入了欧洲现代主义建筑史的教科书中。它所倡导的实用功能主义潮流是在当时德国、奥地利、瑞士等国家的工业联盟经过不断探索和大量经验基础上运用相似组织方式呈现的共同成就。

芭芭住宅区是一个私人业主投资建设，为捷克斯洛伐克制造联盟成员所用的居住区，他们中的很多人（如约瑟夫·戈恰尔和帕维尔·亚纳克等）也是马内斯美术家联盟的成员。项目反映了建筑师（及业主）的民主愿景和社会理想。这块曾经的农地在 20 世纪初被规划为住宅用地，因这里被认为是布拉格最有利于健康的地区，它面朝西风吹来的方向，空气质量极佳。居住区共有 32 栋住宅，建于 1929—1932 年间。从外观上，每个房子都是看似相同的白色体量，但因业主、基地、景观等不同，其表现出不同的个性，共同组成一个小社会。

这些房屋矗立于倾斜的坡地上，抽身于布拉格的历史环境。设计者将这个项目视为展示现代社会理念的一种实验。建筑由钢筋混凝土、玻璃等现代材料建成，但没有采用钢结构及预制构件。设计者认为通过建筑外观上的新材料可以消除历史传统的等级制度。使用平屋顶消除了古典屋顶的坡面形态，使得建筑群具有明显和统一的现代特征，每幢建筑外观都是其内部布局的几何抽象。

居住区整体规划通过一条主要的轴线进行组合，沿着轴线的是宽阔的绿化带（可以作为公共花园）和纵向贯穿的宽大道路。这个时期，汽车的使用和休闲时间的增长都影响了房屋的设计。汽车作为主要的交通工具，车道和车库被纳入建筑规划中。场地的开放性规划与房屋的自由布局重叠，形成了空间的统一连续性。绿化景观及建筑按照现代美学法则被整合到一起，在最小的空间创造出最好的效果。每一栋住宅以及住宅中每个房间，都可以不同程度地被看作其主人的空间肖像的抽象表达。

芭芭住宅区表现了一种多向度的抽象语言，从基地到建筑构件：玻璃转角和玻璃墙的结合、框架空隙、悬挑阳台、自由的柱、露台或雕塑等各种元素被重新组合和表达。与以前传统的大型砖石别墅不同，新的现代别墅开放、连续的空间模糊了空间和功能之间的界限，并由此产生了新的生活方式。

即使以今天的眼光来看，芭芭住宅区仍然是前卫实验建筑的范例。简洁到极致的外形、光滑粉刷的建筑表皮、底

层架空或部分架空功能化的平面、水平长窗、大露台和平屋顶完全背离了布拉格市中心区哥特和巴洛克式建筑为主体的历史传统。在勒·柯布西耶提出的新建筑五点（1926年）和实用功能主义还没有被完全接受的时候，芭芭住宅区的探索对欧洲的现代建筑运动起到了极大的推动作用。

… becoming its own hillside north of the historical city, a new landscape of individuals looks back to the Prague Castle.

In an effort to promote the ideas of their work, a single, privately funded, residential estate was prepared by and for individuals of the Association of Czechoslovak Werkbund, many of whom were also of the Mánes group (namely Josef Gočár and Pavel Janák). They planned to establish a cohesive urban structure of independently-defined component parts, each its own distinguished configuration, using the universal a-historical language of Modern elements. Viewing this moment in time as a possibility to project an advanced Modern world, they built what they believed to be a demonstration of a future society.

Mirroring their democratic vision, the collective achievement of the project was to be the representation of the uniqueness and diversity of shared individualism. Conceptually, each house was the same white volume, yet the particularities which informed them: site, client, views, preferences, etc., marked the range of their owners' individualities, and together composed a small representative society of 32 houses. The individual houses, as with their individual rooms, were all abstracted to varying degrees as spatial portraits of their owners.

The adoption of a new city plan, which permitted new areas for construction on the edges of the city, allowed the project to begin from 1929–1932. The houses were built of reinforced concrete and glass, without steel or prefabrication. The project included the building of infrastructure, a fabric of private roads connecting the field-cluster of separated houses. The car as a source of arrival required driveways and garages to be incorporated into each house.

Open spaces and new technologies, such as the car, and the growth of leisure time, all influenced the dispositions of the houses. The hierarchies of the interiors were reconfigured from those of the previous era, in order to accommodate these new ways of living. Unlike the massive masonry villas from before, the possibility of having an open, continuous space between rooms could blur the limits between space and function, and yield new ways of living — therefore also new citizens.

Such a multi-scalar abstract language, from the site to the component part, used here at this time, was like an alphabet: a variety of elements to be re-used. Many expressed combinations of glass corners and glass walls, framed voids, overhanging balconies, free columns, terraces or sculptural curves. This common vocabulary of the elements of space was fundamental to the concept of the individual and the open society: its language of voids and parts implied that not one was complete on its own. By overlaying the open plan of the site and the free plans of the houses a unified spatial continuum was established.

Conventional hierarchies and histories were erased through the use of essentially historically blank material expressions on the exteriors. In this case, the abandonment of the pitched roof was favoured to achieve the disappearance of the roof altogether. Each house appeared as a geometrical abstraction of its internal arrangement. The houses, set into a sloping site, were without historical context (having only the views to the castle, the views to the river, and the topography of the land).

There were certain contradictions, however, between theory and reality; between ideology and economics; which were the unfortunate shortcomings of the Modernist period: bourgeois houses were too expensive for the average person. Many of the houses have been renovated since, and most have lost their original identities. The few exceptions are the Sukova, the Paličkova and the Janákova Villas.

缪勒别墅
Villa Müller

地址：Nad hradním vodojemem 642/14, 16200 Praha 6, Střešovice
交通：Tram 1, 2, 91, 96 **Stop** Ořechovka（1 分钟步行）
建造时间：1930—1932 年
建筑师：阿道夫·路斯（Adolf Loos）、卡雷尔·勒霍塔（Karel Lhota）

□ 现代主义/MODERNISM

缪勒别墅坐落于布拉格北郊的一个坡地上，是建筑工程公司老板弗朗齐歇克·缪勒的住宅，由出生于今捷克布尔诺的建筑师阿道夫·路斯主要设计。卡雷尔·勒霍塔向业主推荐了路斯作为建筑师，并协助深化内部空间设计，配合现场建造过程。从某种意义上，该建筑的修建本身也成为建造商缪勒的商业名片。

缪勒别墅试图建立一种更深层次的建筑秩序体系，并呈现在建筑外表皮的包裹之中。别墅为钢筋混凝土结构，外表是一个简单厚实的白色立方体，从外部开窗看不出内部异常丰富的空间布置。路斯坚持了德语区对空间的传统理解，即"房间般的空间"。因此，建筑由大小不同、功能各异并独立的房间体积构成；体积之间以开洞的方式连接，以保证彼此的独立性。在此基础上，错层的加入进一步强化了上下界面的分离，楼层之间的半个错层也成为建筑的特点之一。建筑内部空间围绕着中央楼梯布置，楼梯本身成为空间中的雕塑，并形成具有序列感的体验。室内外的模糊性在屋顶的视野中达到高潮，屋顶有一片墙特意朝向布拉格城堡开口，实现了别墅与城市历史的遥相呼应。

别墅内的房间与素净的白色外表面相比，色彩鲜艳，材质华丽。采用众多贵重的材料，诸如桃花心木、橡木、日本纤维墙纸、洞石、大理石、花岗石等进行装饰，部分家具也是定制的。客厅作为最重要的公共部分，以一个开放的L形空间成为社交中心。华贵的墙面石材和具有异域风情的地毯及素净的白色顶面形成对比，并且以这种丰富性构成室内外空间的鲜明反差。客厅空间放置了5个不同风格的椅子，客人可以根据自己的喜好选择。路斯认为"丰富多样的材料和优异的做工不应该仅仅被认为是弥补装饰的不足，而应该被认为它在丰富感上远远超过装饰"。缪勒住宅室内的丰富性，可以修正我们对"装饰即犯罪"的认知——路斯反对的是表面的、肤浅的、违背现代主义的装饰。

路斯自认为他从1904年发展的体积规划理论的最大贡献在于提供了一种在三维中处理建筑空间安排问题的范例。缪勒别墅作为建筑师的代表作，是路斯众多作品中最能充分体现其体积规划理论的一个，并成为那个时代的象征。

该建筑在1997—2000年进行修缮后作为城市博物馆开放。

... anchored into the corner of a Modern suburb, a house with an internal fold contains the Prague Castle within its empty form.

This house has become an icon of a time and oeuvre. It developed a sophisticated new type of spatial unit (greatly admired by architects) by adapting an established tradition of spatial planning, with a progressive socio-spatial theory born after Freud and Einstein. The Villa Müller was originally to be typical and to blend with its traditional context. It became a landmark expression of the individual in a democratic society.

The Villa Müller was originally to be typical and to blend with its more traditional context. The client and owner of a reinforced concrete company, František Müller, agreed to revise this project with the architect Adolf Loos. Upon visiting the site at the corner of a steep slope, Loos proposed his theory of the interior as a spatial concept, known as Raumplan. Through shifts of volume and a subtle asymmetry, the house could be both separated from its surroundings and complementary to their character. Construction took place from 1930 to 1932, using reinforced concrete.

The misreading of Loos' oft-quoted "Ornament and Crime" (in the service of a "theory" of minimalism) can be readjusted by recognising the difference between Revivalist, ornamental rhetoric representing the symbolic, and the experiential excesses of the personal and the private. The interior spaces of the house are considered an embodied extension of the life of the person, and their ornamentation is thus justified (One can not live within a facade).

The rooms of the house are colourful, ornate and decorated with beautiful materials. There are many different materials such as mahogany, fibrous Japanese wall papers, marbles, and custom furnitures. The interior spaces are organised around a central, sometimes bifurcating stair. Each room is an autonomous volume placed upon its own level in space. They are linked through the use of stepped floors into a sequence of experiences. The open form of the living room, with its intersecting L-shaped volumes, is the social centre. The half-stairs between levels are an architectural feature which themselves become sculptural figures in the space.

The social hierarchy was structured throughout the house as internal configurations of rooms experienced as linked volumes. Their form is unrecognisable from the outside. The outer volume appears irrational through a new rationality of public privacy. The few openings placed over the exterior obscure the revelation of order within. While appearing calm, within the constraints of the facade, the interior must be in movement. In an exceptional moment, the southwest corner splits to expose this inner play.

Seen from this corner, no longer cubic but spiral, the mass of the house ruptures and unfolds. Contradictions are exposed from the outside. The doubled line of the cornice (a magnificent ornament) appears at once to recess into and to project from the mass. Continued beyond the roof line, as neither column nor wall, an ambiguous limit of form is crowned in a thin mass of space. The facade becomes visible as a shrouded figure wrapped around a fortified volume. While it may appear mostly blank from the exterior, without readable representation of its internal spaces, the Villa Müller establishes an architecture of a deeper law within, which is revealed in the unresolved closure of its wrapping skin.

As a rational re-writing of Neo-Classicist themes of symmetry and order, the ambiguity

between inside and outside produces on the facade a minimalism which is precisely not style. The windows are a subtle play of language that, while openings, obscure the positions they formed within the internal space layout. Not enough evidence appears to imagine how and where they may be internally resolved. There is no need for ribbon windows to express Modernity, as long as the volumetric mass of the Villa Müller reveals itself to be an unfolding layer. The modern facade becomes form itself, in this meeting point of its dynamic internal figure and its opaque and stoic fronts.

The blurring of inside and outside reaches a climax at the view located on the roof. Looking back from the Romanesque, the building as a mass becomes the figure (its life, its routines, its secrets, etc.) contained within itself. A modern and inverted cubic rotunda, it faces east to the Castle. It reframes itself with the city seen through itself: through the open enclosure of its missing roof. This "framed absence" exposes that there is a perpetual something contained within, and that while it certainty exists it can not be fully known. The Villa Müller is a house for a civilised subconscious.

耶稣圣心教堂
Church of the Most Sacred Heart of Our Lord

地址：Vinohradská 1438/70, 130 00 Praha 3, Vinohrady
交通：Metro A Stop Jiřího z Poděbrad（2 分钟步行）
建造时间：1929—1932 年
建筑师：约热·普列赤涅克 (Jože Plečnik)

现代主义 / MODERNISM

耶稣圣心教堂矗立于维诺赫拉德山坡上的乔治国王广场中,其高耸的形体即使从布拉格城堡也可以远远地被眺望到。1929—1932年间,建筑师约热·普列赤涅克在完成布拉格城堡改造项目后设计建造了这座教堂,因此耶稣圣心教堂与布拉格城堡有着密不可分的联系。在许多方面,耶稣圣心教堂都是布拉格城堡改造项目所展示的时代精神的再现。

这座地标建筑由两个主要部分组成:巨大高耸的钟楼以及水平长方体的教堂正殿。钟楼的外形特异,如同一座变形大钟镶嵌在一个扁平的建筑体上。建筑的两部分有相似的形态构成呼应:下部咖啡色的墙体与上部白色的檐口矩形窗形成一种合适的比例。深色的烧结砖作为饰面材料,其间安插浅色的石块,模拟了"织物"的视觉效果,丰富着外立面的肌理。建筑主要的入口:西立面上镶嵌着三个白色石材门框,大门上方的山墙与钟楼的顶部山墙相呼应。

教堂内部是一个12米高的有着现代工业化特征的平坦大殿,构思来源于诺亚方舟。建筑内墙与外墙一致,也被划分为两个部分:上部是白色的采光天窗,下部红砖墙凹凸有致,形成连续柱的韵律。除了结构性的表达外,教堂里几乎没有其他装饰品。在白色苏马瓦大理石的主祭坛上方,悬挂着镀金基督画像,旁边是六个捷克圣徒。在这种精神性的极简主义中,结构语言就变成了主要特征。钢制屋顶桁架采用网格状的咖啡色木饰面。地面水磨石的圆形装饰图案使人联想到古典教堂大理石的拼花地面。这种内部空间的多维编织(地面、墙面和屋面)使得简洁现代的教堂内部空间有着丰富的细节。教堂还有一个拱形的地下室,通过镶嵌在教堂地面的玻璃砖进行照明采光。

教堂钟楼高42米,基座占地为22米×4.5米。钟楼顶上的十字架高4米,钟面圆窗直径7.5米。从教堂内部狭长的坡道折叠而上可以到达玻璃钟,站在巨大的玻璃钟后可俯瞰布拉格城区和伏尔塔瓦河。钟楼外观随着观看距离和角度不同而呈现出变化,若隐若现,精彩绝伦。时间、城市和人,通过这种玻璃和钢铁组成机械的永恒连接,从而达到了另一种精神性的联通。

耶稣圣心教堂是一座具有现代主义民主精神的殿堂。普列赤涅克在近现代建筑史上被誉为"后现代主义的先知"。他在这幢建于20世纪二三十年代的建筑中所体现的装饰主义有着埃及神庙和早期基督教圣殿的意象,已经形成了与20世纪80年代的后现代主义相似的探索。美国建筑师弗兰克·盖里来这里参观教堂后戏称:"没想到迈克尔·格雷夫斯比我先到了布拉格。"

… in the central park of a new neighbourhood, a memory and a vision to be seen from the Prague Castle.

City, heaven and earth all intersect in this fusion of cosmological and topological grounds. In many ways, the building is the embodiment of the spirit which emanated from the Castle at this moment in time. It also resolves a long-played out conflict in Prague between the city and the Church.

In the crossing of altar and clock tower — Church and city — in the Church of the Most Sacred Heart of Our Lord, the church and the tower are one. The coupling could avoid the legal complications of both St Vitus and St Nicholas, as the clocktower is accessed from within.

The area around the square rapidly grew when it was incorporated into Prague in 1922. Construction by Jože Plečník (also builder of the Prague Castle gardens and courtyards and a member of Mánes Union) took place between 1929 and 1932 alongside the growth of the new neighbourhood. It is composed of two major parts: its Modern church nave which is placed into the park and the clocktower which is visible from the Castle. It was a temple for a modern and industrialised democratic people, and it marked the new expanse over the horizon as it advanced and reinvented a tradition. Its noble character was enhanced with an ancient Greek iconography of triumphant wisdom emerging from within.

The church nave is contained within a massive and unfolding layer of "stone fur". Its upper edges fold outwards into public space, as the temple of enlightenment rises out from within. The cornice recalls an expressive Cubist break as well as the playful mannerisms of the roof at Schwarzenberg Palace. The regal, dark brick patterned fur appears to unfurl an enlightened surge of symbolic stone upwards. This expression of an enlightening form opening up, combined with the textural pattern of the bricks, amplifies its eruptive visual effect.

The west facade is pierced by three white stone portals. The pediment above this three-portal entrance echoes the top of the clocktower. There is an equal measure between the length of the temple and the height of the tower and each has equivalent, complementary value. The flat roof above the church nave is not perceived, as it is surpassed by the imagery of the rising temple top. The roof of the church nave, with its horizontal steel truss grid revealed within, is its own hidden horizon. And in an inverted echo, the pediment of the clocktower, as seen from the Castle stands upon the horizon of the city.

The gridded and reverberating pattern across the facades echoes the radial character of the nave's interior space. The interior volume is an immense, vast, industrial-scale hall 13 metres tall. It is a rectilinear volume with a clear division of three symbolic levels. The high and low levels (the heavens and the earth) of the interior space are woven together by a third interwoven element: the white space between body and spirit. A white, 2-metre-high band of alternating strips of walls and windows is recessed, slightly, behind the brick mass. It appears to rise up from within the body of the Church, to become both structure and light.

The construction assembly becomes a symbolic figure within the church's spiritual-technological expression. The dense grid of steel roof trusses are clad in wooden facing, which appear as if linking threads in a woven fabric overhead. The cast stone ground, with its patterns of radiant circles, suggests the infinite even as its edges meet the brick walls. This multi-dimensional weaving of the spatial fabrics of the interior — an uninterrupted open space of contiguous isometric frames — leads the eye ritualistically both horizontally and upwards at the same time.

There is access to the clock tower from within the church nave, though it does not disturb the interior space plan. The tower stands 42 metres tall with its clock 7.5 metres in diameter. The internal ramp structure folds up above the roof of the church to look over the Modern world. Time, city and man are linked through this eternal machine of glass and steel. The clocktower within the church as a register of all time and a witness of all of mankind's achievements is also a source of light above the city.

共产主义时期

1948—1968—1989

或许这个时期始于 1945 年苏联军队解放布拉格,或始于 1948 年 2 月共产党人夺取政权,也可能始于 1968 年的布拉格之春或 1989 年天鹅绒革命前的任何一年。这个时期充斥着各种冲突和斗争,对如何统一社会出现了各种不同愿景,种种社会思潮的纷争势头高涨。

1948 年,捷克斯洛伐克社会主义共和国成立,并成为东欧社会主义阵营的一员。尽管历史上没有更深地联系,但莫斯科从此对其产生重要的影响。事实上,数个世纪以来布拉格一直处在维也纳的影响范围内,深层次的文化和历史纽带更多地将布拉格与维也纳的政治身份紧密地融合在一起。而此时,布拉格却处于"东"方,维也纳处于"西"方。

此后,计划经济开始实施,自由市场经济被取消,投资和生产均由国家控制。此时,社会价值观崇尚集体主义,这种文化背景下便是多样性的丧失。美学和创造力总是让位于对已有标准规范的遵循,社会空间模式使得个性受到压制,社会也日趋保守。

尽管布拉格人并不喜欢,苏联式的社会主义现实风格还是被引入。这些刻板的构筑物因其巨大的尺度在布拉格形成了一种另类的城市景观。新的基础设施比如地铁的建设加快了城市的发展,一座座运用现代技术和材料建造的大型建筑也被植入这座古城中,但有时是以牺牲城市建筑遗产为代价。社会主义政权在第二次世界大战后为人们提供低成本、紧凑简单的房屋,中央计划控制生产出大量均质化的建筑产品。在整个 20 世纪 70 年代,捷克斯洛伐克一共造了大约 100 万套外形简单,采用预制构件的社会主义住宅,解决了居住的基本问题。尽管这个时期缺乏自由开放的精神,但两次大战期间形成的现代主义传统继续在捷克斯洛伐克起到作用,现代主义还是被带入到这些集合住宅中。在布拉格郊区可以看到这些社会主义时期的连片住宅,形成另一种城市历史图景。

在共产主义时期,自上而下的政府管制和自下而上的民间反抗斗争之间暗潮涌动。1968 年的布拉格之春,因苏联军队入侵被扑灭了政治自由化。这个时代

共产主义时期的布拉格
The Communist City

因此被划分为两个时期：1968年之前，社会主义现实主义风格在20世纪50年代初期被植入，这种纪念碑式的古典主义与捷克的建筑传统不同，并很快在捷克斯洛伐克获得了"sorela"（修女）的绰号。1958年捷克斯洛伐克"EXPO58"馆在布鲁塞尔世博会上大获成功，"EXPO58"以完全非历史化的方式构思，建筑师从第二次世界大战前的捷克现代主义的传统中寻找灵感。从此，建筑唯一的社会主义特征就是在一个自称社会主义的社会中建造。1968年之后，粗野主义和高技的风格占据主流：暴露的钢结构、强调坚实体量和构造、具有雕塑形态、

混凝土板
Concrete Panels

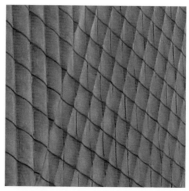

刻花玻璃砖
Cut Glass Tiles

钢制立面
Steel Facades

厚重的面层
Massive Cladding

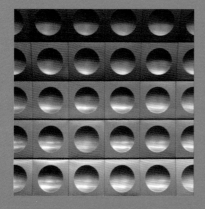

金属瓦
Metal Tiles

抽象象征主义
Abstracted Symbolism

偏爱混凝土等做法，在 20 世纪六七十年代的布拉格（包括整个欧洲）都很流行。新功能主义在 20 世纪 80 年代初成为建筑发展的新趋势。在东西方铁幕隔离中不乏反映时代精神的建筑精品，布拉格在此期间的丰富多样不亚于西欧，体现出捷克斯洛伐克强大建筑传统的延续和在管制制度下的应变。

从共产主义初期苏联风格的国际酒店，到共产主义末期飞船形象的日什科夫电视塔：新古典专制主义和新未来专制主义就像在书中一个章节头尾放置的两张书签，各据一边。

以下这些建筑可以被视为意识形态单极化的共产主义时期的一系列发展和风格碰撞。

THE COMMUNIST TIME

1948—1968—1989

...perhaps it was in 1945 with the eviction of Czechoslovak Germans from the country. Or perhaps in February 1948 when the Czechoslovak Communist party usurped political power and protesting students were silenced by force, when the Modernist vision of multipolarity and diversity ended. Perhaps throughout the spring of 1968, or in 1953 or in 1977, or in any year until 1989, the mixing of tensions and contradictions, between competing visions of a reunified society, were at their peak.

With the Nazi German takeover between 1938 and 1945, the open structures of the First Republic came to an abrupt close. The powers of the Church and the Crown which had been limited during the First Republic were diminished under the Communist takeover in 1948, as the new State assumed all the roles of Crown, Church, and Market, in the name of the People and the Land.

Without having deep historical connections, the land nonetheless became part of a larger union, the Warsaw Pact. While Prague was under centuries of back-and-forth rule with Vienna their deep cultural, linguistic and historical ties fused together their political identities. Vienna, was now "West" and Prague "East".

The new regime was at odds with the Humanistic philosophies of the Modernist time, and with the closure of the free and open ground plan, there followed an oppression of diversity. The new regime, however, embraced the technological progress advanced by Modernity. At the expense of heritage, infrastructure was built. A heavy architecture was now made in modern technology, glass, concrete and steel. Imposing steel constructions were inserted into the stone city. Singular control over the markets yielded the mass production of uniformity. The new art of icons and symbolisms of a State-centric spirit subverted the whiteness of the Modernist ethos — with a clear hierarchy of political power and visual representation not seen in Prague since Romanesque time. The sacred form of architecture was reserved for the appropriate temples and palaces of the State.

Modernism was adapted for its economic efficiency in an ambition to build a counterpoint to the historical city; to provide "minimalist" houses after the war; and to overcome the "unnecessary" embellishments and excesses of

28. 国际饭店
Hotel International

29. 斯拉夫本笃会修道院（修复）
Emauzy Convent

30. 联邦议会大厦
Federal Assembly Building

31. Máj 百货公司
Máj Department Store

32. 布拉格会议中心
Prague Congress Centre

33. 新舞台
New Scene of National Theatre

34. 日什科夫电视塔
Žižkov Television Tower

the images of the past. A new society was produced without the driving-force of mercantilism or the individualism of the modern Humanist. Modernism was brought to the collective house, albeit without its free and open spirit. Nevertheless, memory and heritage in Prague remained.

Despite apparent closure after the war, the Communist time still inherited oppositions and inner tensions. Even while greatly disempowered, there was a struggle to not lose the identity of the pre-war periods. The explicit confrontation between the top-down and bottom-up forces in the society at this time can be put into two periods: before and after the "Prague Spring" in 1968.

After the Prague Spring came a re-occupation of space by force. A violent disruption of the city fabric was enacted by the establishment of new symbolic forms (architectural tanks). In the post-1968 period, a tension developed between historical demolition and the imposition of a high-tech style (a renewal by force). A sustained opposition was continually at play throughout this otherwise autocratic and deterministic period. Ideas from past times were recycled, and occasionally reappeared.

In retrospect, a certain Baroque extravagance in the symbolic buildings built after 1968, reveals a prolonged moment of tension within and against the absolutist yet chaotic order. The Communist time, like the times before, absorbed the oppositional surges into its own form. The architecture was at times historically sensitive, at times brutal, and often both at once. The denial of the past provoked a necessity to invent the future, and conflicts were sublimated into symbols in increasingly technocratic and hyperbolic ways. Ideological teleology, with its promise of a future of technological advance, despite internal contradictions, was re-established as in the previous Imperial times.

From the monumental, yet undermined, International Hotel at the beginning of the Communist period, to the grounded rocket of the Žižkov Television Tower at its end, new Neo-Classicist absolutism and a Neo-Futuristic absolutism stand on either side — like two bookmarks of a chapter.

The following examples can be considered as a sequence of developments and stylistic collisions through the ideologically polarised Communist time.

国际饭店
Hotel International

地址：Koulova 1501/15, 160 45 Praha 6, Dejvice
交通：Tram 8, 18 Stop Nádraží Podbaba （2 分钟步行）
建造时间：1952—1956 年
建筑师：弗朗齐歇克·耶扎贝克（František Jeřábek）、军事设计院
　　　　(the Military Design Institute)

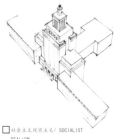

☐ 社会主义现实主义/ SOCIALIST REALISM

国际饭店坐落在布拉格城堡与芭芭住宅区之间,修建之初是为了接待来自华沙条约国特别是苏联的军队客人。作为"不请自来的礼物",其建筑风格可以追溯到在莫斯科建造,被斯大林看作可以与纽约帝国大厦相媲美的苏维埃宫等七姐妹楼。这种建筑风格在第二次世界大战后的社会主义国家盛行,华沙也有一个苏联在第二次世界大战后馈赠的建筑(文化科学宫),但并不受当地人的欢迎。

国际饭店被布拉格人有意安放在远离伏尔塔瓦河和城堡的郊区低地,这使得原本88米高的建筑在古典的城市空间中不那么突兀。此处也是从北面进入古城的开端。尽管如此,它仍然与当地景观要素形成冲突。与现代主义建筑轻盈自由的空间形态相对,这个坚实的构筑物是一种强权和力量的展示,表现出与捷克先前抽象的、个性的建筑样式在物理空间以及意识形态上的对立。

国际饭店的建筑内部结构清晰地反映在外部形态上。建筑呈中间塔楼,两边侧翼对称布置,有278间客房及套间,配套会议厅、宴会厅和咖啡厅、冬季温室和观景台能够满足各种活动使用。因建筑位于斜坡上,使得15层高塔被进一步凸显。另外,建筑立面富有节奏感的阶梯式收分更展现其垂直性。

建筑装修奢华,主入口上方是三组社会主义写实风格的高浮雕,代表着战后人们对胜利的庆祝。深米黄色、粗糙纹理和红色拉毛粉饰让人想起布拉格城堡边的另一个竞争者:施瓦岑贝格宫。拉毛粉饰的细节并没有采用感性抽象的文艺复兴样式,而是通过社会主义写实风格的肖像和图案来表示。华丽的室内则采用大理石、金属构件、马赛克壁画、挂毯以及其他艺术品进行装饰。

早期共产主义的影响在这座城市还留下了其他印记。1950—1954年莱特纳公园修建了世界上最大的斯大林纪念碑,这个20米高的花岗石巨构直戳天际,支配着布拉格老城的天际线。在斯大林去世后,各个国家开始实施"去斯大林"化,这座纪念碑也在1962年被爆破拆毁,此类建筑风格在中东欧国家也逐渐被废弃。

… it stands, isolated in the midst, between the Prague Castle and the Baba Estate.

Prague was mostly undamaged by the war, and did not have the open space in the city centre for such a massive construction. The history of this type of building can be traced back to the unbuilt Palace of the Soviets in central Moscow in 1931, which was conceived by Stalin as a rival to the Empire State Building in New York City. One of these types was built in Warsaw, as a Soviet gift after the war. While this tower symbolised a post-war unity of the central European lands within the Warsaw Pact, it also symbolised an oppressive presence.

Approached from the north, it marks the beginning of the city. However, from the point of view of the city centre it is far outside, unseen. To some degree, it can be viewed as an alternate Castle. But it is placed behind and beneath the Castle. Located far down the river, at the northern edge of the city, despite its 88 metres in height, its impact and power as a rival authority is diminished. Nevertheless, it is a foreign imposition within the landscape, originally intended to house foreign military officers. Unlike the Church of the Most Sacred Heart, the Hotel International projected its red star (now clad in dark metal panels) over the entire horizon. It stands — as both building and political act — in the middle of the shared lines of sight between the Baba Estate and the Castle. Its monolithic form is a topological and a cosmological rupture of the mutual vision between the mythic foundation of the city and the metaphorical ambition of a democratic republic.

In opposition to the free and light spaces of the Modernists, this large tower was an expression of power and solidity, mass and strength. It demonstrated a physical and ideological opposition to the abstraction, horizontality and individuality of the early Modern time. The prominence of the tall, 15-storey tower was further accentuated as it stands on a steep slope. After preparations from 1949 to 1951, it was built between 1952 and 1956 on Družba Square in reinforced concrete and masonry, by František Jeřábek and the Military Design Institute.

It is a complex of four buildings (a tower with two six-storey side wings and a four-storey courtyard wing) filled with social activities. The internal organisation of the Hotel International follows the hierarchy of a centralized vertical axis, like the Municipal House, a rival symbol of authority. Its symmetrical layout, with an entrance hall, club rooms, and 100 rooms and suites, is topped with a restaurant and cafe, a winter garden and a lookout deck. This internal hierarchy is explicitly reflected in the exterior. Its rhythmic stepped pattern directs attention vertically.

The building was finished in rich decorations. Above the three-part main entrance are three sculptural entablatures of figures in high relief, in the style of socialist realism, representing the peoples' triumphs and celebrations after the war. Dark beige, textured stucco and red sgraffito carvings on the facade, recall a second rival of the Castle, the Schwarzenberg Palace. The sgraffito details, however, were represented in icons and symbols, and without the perceptual abstraction of the Renaissance. The interior was also highly decorative, with exceptional marble and ornamental metal details, mosaics, frescoes and tapestries, as well as other artistic works. The building now serves as a luxurious hotel.

In these early years after the war other marks on the city were made in this architectural style. A monument to the leadership of Stalin was completed on Letná Hill concurrently, from 1950 to 1954. This landmark was a 15.5-metre-tall monument in solid marble and it dominated the Old Town skyline. And after his death in 1953, it was demolished (today it is a red metronome) and the Socialist Realist style was abandoned.

斯拉夫本笃会修道院（修复）
Emauzy Convent

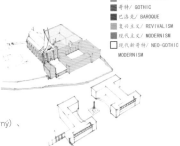

- 罗马风/ ROMANESQUE
- 哥特/ GOTHIC
- 巴洛克/ BAROQUE
- 复兴主义/ REVIVALISM
- 现代主义/ MODERNISM
- 现代新哥特/ NEO-GOTHIC MODERNISM

地址： Vyšehradská 49/320, 128 00 Praha 2, Nové Město
交通： Metro B **Stop** Karlovo náměstí（7 分钟步行）
建造时间： 1965—1968 年
建筑师： 弗朗齐歇克·马里亚·切尔尼（František Maria Černý）、
弗拉迪米尔·坎贝尔斯基（Vladimir Kambersky）

斯拉夫本笃会修道院坐落在布拉格城堡和高堡之间的一座小山丘上，从布拉格城堡、高堡以及伏尔塔瓦河畔都能看到其独特抽象的交叉双抛物线塔楼。作为查理斯四世新城的一部分，斯拉夫本笃会修道院创立于1348年，并成为学习艺术和斯拉夫文学的中心。在胡斯战争、三十年战争以及巴洛克时代复杂的改革期，塔楼都幸免于难，但在1945年第二次世界大战结束时，塔楼被炸毁。这是布拉格历史城区少数几个被第二次世界大战破坏的建筑中最具代表性的一个。

1965—1968年间，修道院的塔楼被作为新技术成就的象征重建，成为一个联系过去、面向未来、修复战后创伤的建筑（当时的意图是将教堂改建成音乐厅，但后来由于技术和经济原因无法实现，在20世纪90年代，它再次成为一座教堂）。新建的塔楼没有复原原有的形式，采用了与当时要求民主开放的国家局势相呼应的现代抽象语言。与同时代自上而下强赠予的社会主义现实主义风格的国际饭店不同，也不像莱特拉公园上显示强权的、世界上最大的斯大林纪念碑（1962年被炸毁），新建部分植根在自身历史的沉淀之上，并且运用当地的材料和新的技术来勾勒出自身的形式。它与较自由的查理四世和托马斯·加里格·马萨里克时代呼应：具有现代的精神、与历史联系并展望未来。在1968年布拉格之春前的几年，人们相信民主及改革是有可能被重建的。因此，这座建筑就像是那个时期的象征，通过现代建筑语言实现理想，并与这座城市原本的哥特根基所联系。

在新的规划中，修道院被安置在一个统一的中心轴线上，这条轴线在更大的构图上包括了河滨广场和现代风格的牧师居住建筑群。项目不仅要对塔楼重建，而且还必须在不新增与这一综合体冲突的建筑的前提下，对缺失部分进行修补。早期的哥特钟塔在巴洛克时期用哥特盛期风格再造，之后在19世纪又被用新哥特风格重建。而现在，塔楼在此被重构成为一个抽象图形：交叉双抛物线的形式。这样塔楼不会过于形成中心也不会过于独立，从而达到一种平衡。

采用双塔的形态既出于结构目的，同时也创造出间隙空间。这一造型中所形成的间隙使得建筑形态及结构更加轻盈，并且也象征着开放激进的姿态。整个塔有25米高，但塔尖仅有40厘米厚。产自捷克北部工业区的透明有机玻璃被当作模板，用于塔体混凝土的浇灌。先进的工程技术在这个时代是被鼓励的，通过技术表现，现代主义的抽象形式得以回归。

斯拉夫本笃会修道院受14世纪胡斯运动和17世纪天主教以及20世纪后期现代主义的共同影响。这些不同风格和表现的对比和叠加，共同构成了一个统一和多层次的综合体。朴实简单的建筑外观和庭院与用壁画、陈设和雕塑装饰的室内空间形成鲜明对比。新的塔顶仅赋予最少的装饰，混凝土模板的残余线条以及塔顶包裹的4米高的镀金尖顶被凸显出来。所有这些历史痕迹都成为一个平衡的整体，来阐释这个场所作为历史的象征和在城市中的重要性。

与修道院临近的布拉格规划与发展研究所和建筑与都市规划中心都是同时期的作品，深色钢结构体现出那个时代的技术特征，由卡雷尔·普拉格设计，也值得一看。

…upon the walls of a Gothic monastery along the river, on a hill between Prague Castle and Vyšehrad, a recollection.

Unlike the top-down imposition of the International Hotel, or the monument to Stalin, this structure is built upon an historical settlement and draws its own form from the material upon which it is based. It connects back to the more free periods of Charles IV and Tomáš Garrigue Masaryk, with a spirit of the present bound to the history of the land. In these years before 1968, there was the possibility to once again reform and recompose society. The construction was thus a symbolic moment. By looking inward in order to look forward, the building reconnects through the modern language into the Gothic roots of the city.

The Emauzy Convent was founded in 1348, as part of the New Town of Charles IV upon an earlier Romanesque cloister. The old gothic towers could be seen from both the Castle and Vyšehrad. The towers were spared destruction during the Hussite wars and the 30 Years' War. They were redesigned during the Baroque-era reformation of the complex and again in the 19th century. In 1945, at the end of the Second World War they were bombed and destroyed. While only one of few ruins in the centre of Prague, it was one of the most symbolic.

Between 1965 and 1968, the towers were re-built to be an icon of technological achievement, a reconnection with the past, and a resolution of the traumatic post-war history. Built by architect František Maria Černý, its expression is a return to modernist abstract forms. Its reconstruction corresponds with a broader interest to reconstruct cultural heritage and was coordinated by the State Institute for Renovation of Historic Towns and Buildings. (At the time, the intention was to make the church into a concert hall, but this later proved unachievable for technical and economic reasons and in the 1990s, again, it became a church.)

The church towers are part of an ensemble that is more complex than a single building. They are aligned on a central axis which includes the Modernist-era complex of ministerial buildings and the riverfront square within its larger composition. The reconstruction thus innovated a solution for rebuilding the towers while resolving the larger scale composition without overwhelming the whole with a new dominant. The towers are recomposed into a unique single double figure in such a way that despite its centrality and scale it would not be overpowering but in delicate balance. The doubled tower is formed for structural purposes but the space framed within is also important. It keeps the perception of the structure lighter and also symbolises openness itself.

Only 40cm thick despite being 25m tall, the towers, which look like thin shells, are curved walls. Transparent plexiglass, crafted in the northern industrial region, was used as a formwork for pouring the concrete. The traces of the casting process are expressed as geometrical ornamentation on the exposed concrete surface. This advanced engineering was encouraged in support of technological progress, but its textured abstraction also referenced back to the Modernist sensibilities.

The complex arrangement of the Emauzy Convent is a combination of 14th century Hussite and 17th century Catholic influence with the late 20th century secularism. This contrast and overlay of different styles and expressions together form the unified and multi-layered complex. The stark and simple appearance of the buildings and courtyards contrasts with the decorative interiors of murals, ornament and sculptures. The modern towers are capped with only minimal ornamentation, in this case the residual lines of the concrete formwork, and the abstracted golden tips of the spires. That all of these histories become a balanced whole explains the significance of this site in the city as a symbol of history.

The Institute of Planning and Development (IPR) and the Center for Architecture and Metropolitan Planning (CAMP) can be found in the neighbouring ensemble of black steel buildings designed by Karel Prager at the same moment in time.

联邦议会大厦
Federal Assembly Building

地址： Vinohradská 52/1, 110 00 Praha 1, Vinohrady
交通： Metro A, C Stop Muzeum（1 分钟步行）
建造时间： 1966—1973 年
建筑师： 卡雷尔·普拉格（Karel Prager）、伊日·卡德扎贝克
（Jiří Kadeřabek）、伊日·阿尔布雷赫特（Jiří Albrecht）

■ 现代主义/ MODERNISM
□ 新构成粗野主义/ NEO-CONSTRUCTIVIST BRUTALISM

建于1966—1973年的前捷克斯洛伐克联邦议会大厦，坐落于1885年建造的庄严的国家博物馆和1888年建造的华丽的国家歌剧院（原新日耳曼歌剧院）之间，其现代而庞大的躯体架设在1938年建成的前证券交易所大楼上。这个由卡雷尔·普拉格（也是新舞台的设计者）领衔设计的项目在当时是市中心最昂贵，也是最具争议的建筑之一。

联邦议会大厦的建设体现了捷克斯洛伐克在1968年（布拉格之春）前后的社会剧烈变化。此时的布拉格处于一个很微妙的境地，要继承历史但同时要与过去保持一定的距离，超越了过去但又召唤着复兴。因此，项目以前卫的姿态展示了在哥特历史城区所聚合的一种新型力量的存在。

这是一个多向度的建筑，通过新的材料、构图、比例及高度聚拢并整合了场地周边环境和形态。建筑与毗邻的国家博物馆同高，从空中看，它们互为镜像，其相似的大体量将城市空间拉伸开来。建筑的几何形体部分悬浮在空中，部分架设在历史建筑上，通过其相互交织的体量以期与城市在多个层面上进行对话。证券交易所的石质外墙为这座更轻但同样简洁的联邦议会大厦提供了坚实的基础。建筑的上部分是在地面上组装，并使用液压杠杆物理进行抬高。建筑矗立在6个巨柱（其中两个隐藏）上，每根柱子都是一对C形的钢墩，背靠背连接在一起。建筑上部分立面上金属竖肋与原证券交易所的柱廊形成韵律上的呼应。与证券交易所连接的新建筑体保持了与交易所相似比例和颜色的石材，这样在形式和材料上与老建筑产生联系。建筑产生的两个虚空间：一个是面向国家博物馆的入口高耸空间，模糊了建筑与城市的界限；另一个是隐藏、私密的四层内部中厅，视线不透明也无法接近，增加了复杂的建筑体验。普拉格考虑到这座建筑未来可能会发挥其他功能的可能性，因此室内空间设计具有装配性和可变性的特点，使建筑很容易转换为一个文化场所。

直到1992年捷克斯洛伐克分裂时，该建筑一直被用作国家的议会大厦。戈尔巴乔夫、撒切尔夫人、老乔治·布什和伊丽莎白女王等政治人物都曾到访这里。天鹅绒革命后，它作为自由欧洲电台的办公室，2009年该大厦被用于文化功能，并入国家博物馆。2018年，为了庆祝第一共和国（捷克斯洛伐克）的百年诞辰，它与国家博物馆一起重新开放，并且从地下连通到国家博物馆，在另一个层面上两个重要建筑又被重新连接起来。联邦议会大厦的变迁就是现代捷克社会风云变化的缩影。

... within the city, yet above and outside it.

The Federal Assembly Building captures the powerful contradictions of the 1968 era. It is powerful yet ambiguous; while it overpowers the past it is conjuring the spirit of revival. It takes a tenuous position at a critical point in the city, re-collecting everything around it while at the same time keeping a distance. On the one hand dominating, on the other hand complementing — in either case — demonstrating the presence of a new form of power positioned at the upper threshold of the Gothic city.

Within the footprint of the former New Town wall, and within a dense urban fabric is a complex of interlocking volumes and powerful cantilevers. The composition integrates the surrounding structures of the site by superseding them within one new, if dispersed, centralised spatial structure. Between 1966 and 1973, the Federal Assembly Building was built — to be the temporary seat of the Parliament — and during the spring of 1968, it was a construction site. It was constructed by Karel Prager upon the former stock exchange building of 1938 and adjacent to the National Museum of 1885 and the German State Opera of 1888.

It is a building in the round which communicates with its surroundings through materials, reflections, proportions, and heights, to link the city on multiple levels. It responds to the city fabric and appears to form new relations within the landscape. It implies as much void as solid. This is further emphasised at the corners of the volumes, which cantilever into the city, defining large scale open spaces beneath. This multi-faceted, conflicted layering of the forms and their voids subdivides the monumentality into manageable parts.

The crowning volume of the Federal Assembly Building pulls the panoramic space of the city around from the south. This superstructure was built of 4 steel Veriendeel trusses, assembled on ground level and physically raised and rolled into place using hydraulic levers, and stands upon 6 massive columns (two are hidden). Its horizontal form is aligned at the same elevation as the National Museum. Seen from the Old Town Hall, the two appear as a couple, one the extension of the other. Their rhythmic colonnades and similar colour palettes echo together.

The grounding volumes react differently on each side. On the western side, attached directly to the stock exchange building, the new volume is clad in stone to match the colour of its neighbour. Different types and subdivisions of glass reflect the National Museum and present a smaller scale and perceived variety on ground level. On the eastern side, a smaller volume which matches the tone of the German State Opera, is displaced overhead to exaggerate the heaviness of the embedded glass volume in the ground. On the northern side, all these material components in glass, stone and steel are revealed at once as separate and distinct parts.

A towering volume of void carves the primary point of entry into the earth on the southern side facing the National Museum. The structural columns are a dominant figure in this space. Each column is a pair of two C-shaped steel piers joined back to back. Placed as a couple, they stand representatively in a recollection of all monumental gateways of the past. Their abstracted individuality combined with their scale and height of 16 meters, give the columns an image of both humanity and oppression and they recall the mythical guards at the gates of Clam-Gallas Palace.

The most mysterious volume is an elevated exterior space on the fourth level. While a technical consequence of the assembly process and a means to bring daylight within, it is also a metaphor. This open courtyard is, ironically, visible from outside, yet inaccessible from within. It is a paradoxical absence contained within the symbolic crown of the powerful new government building. The space is both a central heart and a total void at once, embedded within the contradictory construct.

After the crushing end to the uprising in Prague, in the early winter of 1969, Jan Palach, a student, immolated himself in protest of the occupying forces. The memory of this act is built as a bulging mound in the ground facing the Federal Assembly Building, at the base of the steps below the National Museum. The Federal Assembly Building was used as the Parliament until the end of 1992, when Czechoslovakia split. After nearly 15 years as the offices of Radio Free Europe/Radio Liberty, the National Museum took it over in 2009. In 2018, in celebration of the centennial of the First Republic, it re-opened. Changes were made to the interior and the facades were renovated. Emphasis was placed on improving the public landscape which now includes a new underground connection into the National Museum building — linking them again on another level.

Máj 百货公司
Máj Department Store

地址：MY, Národní 63/26, 113 89 Praha 1, Nové Město
交通：Metro B **Stop** Národní třída（2 分钟步行）
建造时间：1972—1975 年
建筑师：约翰・艾斯勒（John Eisler）、米罗斯拉夫・马萨克
　　　　（Miroslav Masák）、马丁・赖尼什（Martin Rajniš）

■ 复兴主义/ REVIVALISM
■ 现代主义/ MODERNISM
□ 新功能后现代主义/ NEO-FUNCTIONALIST POSTMODERNISM

Máj 百货公司位于国家大剧院和亚德里亚宫之间,建在地铁 B 线民族大街站的出口上,是 20 世纪 70 年代布拉格地铁工程大型项目建设的一部分。随着地铁的建设和城市社会景观的重建,在历史城区拆除了大量的建筑。Máj 百货公司与共和国广场上的另一个同时新建的 Kotva 百货公司在城市结构和功能中互为镜像。两座百货公司的建造使基地上有价值的历史建筑被拆除,彻底改变了该地区原有的哥特城市肌理,并为历史地段插入了新的比例尺度。

Máj 百货公司的建造象征着这座历史城市进入到一个重点面向未来的新时代,并向当时些许消沉的社会提供了前进的希望。该项目由来自捷克利贝雷茨的 SIAL 建筑公司(由卡雷尔·胡巴切克在 1958 年创立的捷克斯洛伐克著名的建筑设计公司之一)的三名建筑师设计,他们的设计反映了年轻一代对机器美学的兴趣。项目由来自西方资本主义国家:瑞典的一家建筑工程公司施工,薄壳的混凝土板外墙、水平的长窗、巨大的玻璃表面、非对称的建筑元素突出了其新功能主义的特点,建筑施工仅用了一年半的时间,体现了当时高技术的建筑新特征。

这座百货公司建筑并不是通常意义的街区角楼,它通过一个非常特殊的边缘来进行界定:具有现代雕塑感的外表皮反衬了它周围复杂的空间结构。从建筑对面的西北方向回看建筑,城市和建筑空间的组成形态被揭示出来——两种非对称的、对比鲜明的建筑外墙形成不同效果——沿着民族大街的白色立面是伸展的,顶部向内倾斜,以一种优雅的姿态来展示建筑的长度,并将自己与街道的边缘略微拉开一定的距离。在斯帕莱纳大街却变成一个陡峭的切面,坚硬的素混凝土墙悬挂在建筑物外表皮下。

沿着这个垂直的边缘有两个截然不同的空间片段:在首层,中世纪城市的印记被重新刻画成建筑转角处黄色框架的窗户立面上。建筑基座被削减到民族大街

219

上的第二和第三层的高度,加强了街道之间的层级关系,呼应了国家大剧院门廊的比例。同时,Máj百货公司屋顶的弧形轮廓与国家剧院的弧形屋顶相呼应。百货商店与它的相邻建筑高度一致,也与弯曲的街道产生共鸣。

建筑外立面使用悬挂轻质的面板作为其更薄的表面。沿着民族大街,连续的透明玻璃窗和白色薄金属面板交替。沿着斯帕莱纳街,钢结构上悬挂着混凝土板,在沉重和轻盈之间产生了明显的对比。这种方式将立面表现为一层纤细的薄膜,楼层地板和建筑重量都可以被遮盖,薄如纸的外表皮减轻了建筑的厚重感。同时,它极度平整的外表也使得百货公司的展陈布置能够最大化。

虽然Máj百货公司的建成修补了这一城市街角,但是却无法掩盖其背后城市肌理中的空白,其背后的区域几十年来一直处于闲置状态。在2015年,地铁建成近40年后,这一街区终于完工。随着邻近的四重奏商业综合体的建成,形成了与百货商场的对接联系,它们从内部相互连接起来,使街道景观恢复了生机。

… at a critical intersection along the former Old Town wall, a massive void brings the metro into a city gate.

After 1968, the Máj Department store was built into a newly established era in the city, with its focus on "the future". In this normalisation period of the post-uprising context, the promise of progress was provided to the disappointed society, not long ago accustomed to the free and open spirit of the Baťa Palace (not far down the street). The building was part of the larger cultural project of constructing Prague's underground rail system. With the construction of the metro, and the reconstitution of the city's social landscape, a lot of demolition was required.

The metro was built to reinforce the city's triangular urban structure and to connect the new residential periphery into the centre. The building of the Máj Department Store at Národní Třída and Spálená Street took its place in parallel to the construction of its mirror-site, Kotva, at Náměstí Republiky. Complete demolition of the area's original Gothic fabric and valuable buildings was required in both cases. These constructions inserted a new scale and type of function into the historical ground. Máj was realised between 1972 and 1975 by the SIAL Group of Architects, John Eisler, Miroslav Masák and Martin Rajniš, with the Swedish construction company SIAB.

The Máj Department Store marks a new corner, along Národní between the National Theatre and the Adria Palace. It is a place for shopping with a combined institutional and residential character. A compromise was struck between the significance of the department store within the city, and its place within a dense fabric of residential blocks. It is not a typical corner building—it is defined with a very particular edge. This sculptural edge indicates the complex spatial fabric of its surroundings and defines a volumetric landmark that is both contextual and freestanding.

At the opposite corner seen from the Northwest, the play of space, city and form is revealed. Two kinds of atypical, contrasting walls produce this effect. Along Národní, the street wall is angled inwards and stretched in a manneristic gesture to exaggerate its perceived length by setting it slightly apart from the edge of the street. On Spálená Street, an abrupt cut reveals a stark and exposed concrete wall that hangs along the skin of the building overhead.

Along this complex vertical edge are two distinct moments. At ground level, the memory of the medieval city is re-inscribed into the facade with yellow framed windows which wrap around the corner. On Národní, the plinth is cut back to the second and third levels to reinforce the hierarchy between the two streets. This gesture also echoes the proportions of the portico of the National Theatre, as does the curved roof volume overhead. It links with the heights of its neighbours and also resonates with the surrounding curved avenue.

The facade is made of hanging, thin panels which produces the appearance of a lightweight wall surface. Along Národní, continuous windows alternate with strips of glass and thin metal sheets. Along Spálená Street, concrete panels hang from the steel structure, producing an apparent contradiction between heaviness and lightness. In this manner of expressing the facade as a layering of fine, thin membranes, the thick floor slabs and heavy weight of the building could be masked. The paper-thin envelope reduces the massive volume into surfaces. At the same time, its extreme flatness highlights the objective bulk of the Department Store (which anticipates its contents filled to the limit).

While it resolved the city block at the corner, it could not conceal the void within the urban fabric which was created behind it. On the other side of this masked front, remained a missing city block left exposed and unfinished for decades. In 2015, almost 40 years after the construction of the metro, Národní Třída was finally completed with the construction of the adjacent Quadrio building. They are connected together at ground level and return a vitality to the public space at street level.

布拉格会议中心
Prague Congress Centre

地址：5. května 1640/65, 140 21 Praha 4, Nusle
交通：Metro C Stop Vyšehrad（1 分钟步行）
建造时间：1976—1981 年
建筑师：雅罗斯拉夫·迈耶（Jaroslav Mayer）、安东尼·瓦涅克
　　　　（Antonín Vaněk）、约瑟夫·卡尔利克（Josef Karlík）

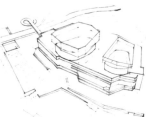

☐ 新功能粗野主义/ NEO-FUCTIONALIST BURTALISM

20世纪70年代新建的地铁C线将布拉格从历史城区延伸到南部的新城区。布拉格会议中心就建在地铁C线高堡站旁的山丘上，由军事设计院的三名建筑师在1976—1981年间完成。这里位置优越，交通便利，是新的城市中央商务区的重要节点，并与布拉格城堡遥相呼应。

布拉格会议中心规模庞大，是捷克最大的文化设施，也是欧洲最大的会议中心之一。建筑地上五层，地下三层，总建筑面积168 000万平方米，建筑有20个厅和50个会议室，总共可容纳近万人活动。其中最大的厅是国会大厅，通高四层，有2764个座位，是一个与美国波士顿音乐厅和英国伦敦皇家阿尔伯特音乐厅齐名的世界级厅堂。

从远处眺望，布拉格会议中心因巨大的建筑体量主宰了历史城区边缘的景观，也成为附近后期建造的摩天大楼的前景。邻近的高层酒店、公路和桥梁形成一片新区，每个新构筑物都是当时一项显著的技术成就，代表了哥特时代城市边缘一种新的城市形态。从形态上看，这是一座庞大的单体不规则建筑，被设计为退台状的结构，充满了力量感，它的体量和外观也具有捷克立体主义的风格特征。布拉格会议中心很好结合了地形并组织了周围环境，室外广场其实是建筑设备及服务层的屋顶，与地铁、高速公路及高堡历史景区交通联系顺达，使得公共文化建筑的大客流可迅速分散。

建筑正立面采用了大面积的玻璃幕墙，用以减轻建筑的体量感，也加强了室内外景观的联系，但咖啡色的玻璃也相对增加了建筑的厚重感。建筑主要出入口位于南侧，首层为门厅、存衣厅、展览厅等，三层交通厅可扩大成各种活动厅，空间流通，可同时举行音乐会、茶话会以及宴会。三层与四层、五层休息厅有共享空间，空间的相互穿插渗透，并配合以现代抽象雕塑，形成生动活泼的气氛。五层有全景餐厅，餐厅开敞的大窗视野辽阔，可鸟瞰窗外远处布拉格城堡的景色。北侧首层为贵宾出入口。建筑地下层为车库、厨房，以及建筑设备系统等用房。

为了体现国家的威严，会议中心的室内设计整体开阔而大气。内部装饰色彩明快并注重细节，主要运用大理石、皮革、木材、黄铜、玻璃、塑料等材料，并融合了艺术家的雕塑、挂毯、绘画、水晶作品等进行装饰，极具捷克特色。

这座建筑之前是捷克斯洛伐克共产党举办代表大会的地点。目前用于举办会议、展览、音乐会和表演等各种公共活动。2018年举办了两次设计竞赛：一个是将建筑南侧画廊空间扩建出一个新的独立体量。另一个是对北部景观区域的更新。这些都会在不久的将来实现。

... outside the New Town city walls, a counterpoint to both Prague Castle and Vyšehrad — it is its own hill.

An enormous volume and scale, like an echo of Prague castle, was set alongside Vyšehrad, to form its greater counterpart. It is a free-standing enclave, like St Martin's Rotunda, but it is a private enclosure for a different purpose and at a much larger scale. It was to be the new representative meeting spaces of the Communist party in Prague.

Anchored into a new metro station, it extends the historical centre southwards into a new centre outside the former city walls. Conceived as the foreground of a new modern district of skyscrapers, it was to be a symbol of a new foundation. An adjacent hotel tower, a highway and a bridge connection were part of the new construction complex. The integration of all these elements, each a remarkable technological achievement, represented a new urban form of progress at the edge of the Gothic-era city. The project was realised, as the Palace of Culture, between 1976 and 1981 by architects Jaroslav Mayer, Antonín Vánek and Josef Karlík, within the Military Design Institute.

The Prague Congress Centre rests upon its own plinth, composed of multiple terraced grounds. The greenery of the terraced landscape is in fact not land but roofed service spaces. While the plan of the landscape makes a connection into the natural setting of Vyšehrad, the mechanical equipment and concrete structures emerging from below reveal it to be artificial ground. It represents a kind of fortification, demonstrating great power as it projects over the horizon.

The building is both monolithic and varied. It contains large and small volumes which are woven together with circulation and gathering spaces to produce an irregular, asymmetrical form. This diversity is reflected with two materials on the facade. A curtain wall of reinforced concrete panels, finished with a white cement surface is combined with plates of tinted brown and grey glass. Its large glazed walls are subdivided and proportioned to reflect the multiplicity of spaces within. The two enormous auditoria are free-standing figures within this expansive open space wrapped in heavy, floating walls. The flowing, three-level foyer frames a magnificent panoramic view.

Excess had its shortcomings. The structure is made of steel and due to excess production was reinforced with extra steel, making powerfully solid floors and strong horizontal lines on the facade. The large auditoria are revealed on the exterior by exceeding the frame of the facade and projecting beyond the roofline. With a bursting appearance similar to the affects of Rondo-Cubism, this upper mass however also disrupts many views. The upper volume has become an enormous wall for advertisements which are seen even from the Castle. At such a privileged position within the structure of the city, upon its own elevated landscape, its massive character and dark glass present more a wall than a gateway. Perhaps had the glass been more transparent, the internal forms and generous multi-storied void spaces would have dematerialised the overall mass into a lightweight shell containing sculpted figures in high relief and people in motion.

The vast interior spaces are made for thousands of visitors who pass through the entrances, cloakrooms, restaurants and foyers. Offices and dressing rooms for artists are on the south side and there are three basements for cars and buses, technical facilities, a power plant including mechanical systems, storerooms, an industrial kitchen, and state-of-the-art electronics for the auditoria. The luxurious interiors are filled with bright colours and details. Singular features were installed from the lamps to their custom bulbs. The interiors, typical of such prestigious buildings, are finished with marble, leather, wood, brass, plastic and stone. Installed works of art include engravings, tapestries, paintings, sculptures and reliefs.

The building is currently used for public events, conferences, and concerts — yet questions of its external appearance have been raised. In 2018, two competitions were held. The first, for the extension of gallery spaces on the southern side within a new, freestanding volume and the second, for the renewal of the landscape areas on the northern side. These developments should be realised in the near future.

新舞台
New Scene of National Theatre

地址：Národní 1393/4, 110 00 Praha 1, Nové Město
交通：Tram 1, 2, 9, 17, 18, 22, 25, 41, 93, 97, 98, 99 Stop Národní divadlo
（2分钟步行）
建造时间：1977—1983年
建筑师：卡雷尔·普拉格（Karel Prager）、帕维尔·库普卡（Pavel Kupka）

■ 新文艺复兴 / NEO-RENAISSANCE
□ 粗野后现代主义 / BRUTALIST POSTMODERNISM

作为老城区最重要、最独特的现代建筑，并作为国家大剧院功能的扩展，新舞台以一个完全现代的形象矗立在古典的国家大剧院旁边，用一种全新的姿态介入历史环境，反映了文化、景观、城市和历史之间的多向度联系。

由于战争对大剧院的损坏和本身功能扩充的需要，从20世纪20年代开始，国家大剧院举办了多次扩建的设计竞赛，先锋建筑师博胡斯拉夫·富赫斯在20世纪50年代初的竞赛中赢得了设计，但他却在建筑建造前的1973年去世。后来国家历史城镇和建筑改造研究所的建筑师帕维尔·库普卡在1976年接手该项目，建造了南翼和东翼。由于国家大剧院决定在民族大街上增加一个新剧场，整个项目又交给了卡雷尔·普拉格（也是联邦议会大厦的建筑师）并完成了民族大街临街最醒目的主建筑。

最终，在民族大街、迪瓦德尼街和奥斯特拉维纳街之间，形成了一个新的城市街区。一面是国家大剧院的古典立面，另三面则由三座不同的现代玻璃建筑围合。整个建筑群内部包括门厅、新舞台、行政、排练和咖啡馆等多种功能，并围合成了一个可以经常举办文化活动的公共广场（现以瓦茨拉夫·哈维尔命名）。

这个文化综合体有双重的城市功能：在城市内部进行交流，使其重新焕发活

力；在城市结构中引入一种新型的公共场所，新舞台和国家大剧院围合的庭院成为新的、开放的空间。同时，不同时代的技艺都再次浮现在这组文化建筑中。

沿着民族大街，卡雷尔·普拉格把新舞台的整体形式划分成两个互补的玻璃体块，以达到与国家大剧院相协调的比例。玻璃体块分为两部分，一部分是直切玻璃板组成的全玻璃幕墙，大面积的玻璃反射出邻近的古老砖石建筑，使之与周围环境的对比变得柔和；另一部分是半透明的凸透玻璃砖，每一块都独立成形，并通过加固件相互连接，然后一起悬挂在钢结构上，组合形成一种编织关系。由于采用的玻璃砖具有半透明的特质，会映射出人在建筑内部行走的模糊影像，为建筑平添了许多细节与趣味。

从建筑外形上，新舞台的玻璃体块的收分呼应着国家大剧院屋顶的四分之一体量的削减。单个的玻璃砖块也再次呼应了国家大剧院外墙切割石材的棱柱形体块。这个建筑外表皮由玻璃艺术家

斯坦尼斯拉夫·利本斯基和亚罗什拉瓦·布里赫托娃设计，玻璃由捷克本土的卡瓦列尔工厂加工，共计4306块，展示出波希米亚精湛的玻璃技艺。

在室内，产自古巴的大理石将柱子、立面及地面包裹，透过外墙玻璃也可以看到室内的亮绿色。暴露的通风系统被设计作为一种雕塑元素，围绕在楼梯上方一个巨大的悬挂玻璃雕塑周围。室内精美的装饰细节包括定制黄铜配件、家具和灯具等，这些都是艺术家协作完成的作品。二楼的咖啡厅是演员、观众和公众聚会场所，新舞台主要演出包括黑光剧在内的现代剧、芭蕾和歌剧等。

三座新建筑与国家大剧院围合而成的瓦茨拉夫·哈维尔广场长50米，宽40米。这个空间不仅可以通达所有建筑，而且还可以渗透进整个城市街区，将中世纪城市肌理重新融入20世纪的城市中，将现在和过去结合在一起。这个广场作为一个新的城市舞台，经常举办各种露天的城市文化活动，它同时又是这个建筑巨大地下空间的屋顶。这个地下结构还有一条隧道连接两个剧院。透过两个剧院之间的空隙，可以看到不远处的马内斯美术馆和会跳舞的房子。

...an extension of the National Theatre, its modern partner, and a mirroring of the historical city around them.

Reflecting a connection between culture, landscape, people and history, the New Scene of the National Theatre has a double urban function: to communicate within the city and to revitalise its fabric. It introduces a new type of public ground within the city structure. While the Federal Assembly Building internalised and detached the courtyard, in the New Scene, the courtyard becomes a new and open forum and a basis for a social and spatial exchange. The arts and crafts of the time, the Neo-Renaissance surroundings, the Gothic infrastructure of the city and hints of modernism re-emerge in this cultural annex as one.

It was built between 1977 and 1983 in two phases, as an extension of the theatre with new administrative and rehearsal rooms and a new stage. The State Institute for Renovation of Historic Towns and Buildings began the demolition of the existing houses on the site including the former administrative building of 1928 with architect Pavel Kupka. Construction of the south and east wings then began. A subsequent decision was made to change the design of the northern wing into a second stage for the theatre. Between 1981 and 1983, Karel Prager (the same architect of the Federal Assembly Building) completed this northern wing on Národní Street with his revision to the facade concept, in cooperation with glassmakers Stanislav Libenský and Jaroslava Brychtová.

The three sides of this new city block, between Národní, Divadelní and Ostrovní streets, are composed of three distinct glass volumes containing the entrance foyer, the new stage, administration and cafe. The massing concept of these forms, with their permeable parterre, frame an open, public ground (now named after Václav Havel). This 40 by 50 meter space, neither courtyard nor square, faces the east facade of the National Theatre. While providing access to the new buildings, it also brings the space of the city through the block. It re-inscribes the scale of the medieval fabric and its passages into a 20th century form. This "unbuilt" space joins together the present and the past, in a provocatively open cube. As a ground level plinth, it is a new civic stage itself. Yet it is in fact the "roof" of a massive underground structure. Beneath 5 levels of basements, there is a tunnel which connects the two theatres.

Along Národní, Karel Prager divided the building into two complementary glass parts to achieve better proportions with the National Theatre. There are two types of glass. Fully glazed curtain walls with straight cut plate glass combine with a hanging translucent skin of convex block lenses. The glass blocks are a woven fabric with each specifically formed to reinforce the next, and together they hang as a textile from steel construction. The large sheets of glass reflect the masonry facades across the street to soften the contrast between them.

The form of the individual glass block echoes that of the separate glass buildings —and both recall the bevelled edge of the National Theatre roof. The glass blocks also reflect the prismatic massing of cut stone, particular to the Neo-Renaissance style. Their chiseled quality, moreover, enhances their properties of translucency and light. An etched pattern slips over the northwestern corner of the facade to reveal a mysterious spatial figure sliced within the surface, adding nuance and a subtle layer of motion blur. It is especially pronounced during sunsets.

Within the interior, Cuban marble is everywhere. The bright green colour is seen through the glass facades in a reflective luminescent hue. Even the columns and the edges of the floor slabs are wrapped entirely in this marble. The use of the vibrant material continues across the floors and walls of the interior. There is a continuous, vertical foyer space between floors. At the heart of this open foyer, in a café on the second level, is a meeting place for actors, the audience and the public. The details within are a collaborative work among artists, including works of custom brass, furniture, and light fixtures. Even the exposed ventilation system is treated as a sculptural element around a large hanging figure of glass sculpture above the stairs.

Today, there are performances of Laterna Magika and various exhibitions, activities and events in the public space. Through the gap between the two theatres, both Mánes Union and the Dancing House can be seen.

日什科夫电视塔
Žižkov Television Tower

地址：Mahlerovy sady 1, 130 00 Praha 3, Žižkov
交通：Tram 1, 2, 5, 7, 9, 11, 15, 25, 26, 37, 91, 92, 95, 98 **Stop** Olšanské náměstí（9 分钟步行）
建造时间：1985—1991 年
建筑师：瓦茨拉夫·奥利茨基（Václav Aulický）
结构工程师：伊日·科扎克（Jiří Kozák）

■ 新古典主义/ NEO-CLASSICAL
□ 新未来主义/ NEO-FUTURISM
■ 当代/ CONTEMPORARY

从远处看，日什科夫电视塔矗立在布拉格的高处，从古典的城市空间中凸显出来。作为捷克最高的建筑，这座216米高的电视塔建成后立即成为布拉格最醒目的建筑。同时它又因超现实的未来飞船一样的奇异外观，一直被当作"异类"，饱受争议。

电视塔是一个强劲而创新的钢结构建筑，重达11 800吨，由俄斯特拉发的一家工业工程公司建造。建筑通过三个巨大的双层钢套立柱矗立在钢筋混凝土基础上，钢套管之间填充30厘米厚的钢筋混凝土。建筑基础直径30米，埋深15米，塔高216米。主管直径为6.4米，其余为直径4.8米，高134米。这三个结构套管内设电梯和楼梯，三座立柱支撑着三个不同高度的平层：62米处的餐厅、93米处的观景台，以及130米处的服务平台。因其建设难度，该建筑在施工过程中产生了许多专利。

建造三个塔柱是为了削弱建筑外观的体量感和减少对城市轮廓的破坏。从技术上讲，它会更稳定，维修费用也会降低。也许只有站在建筑底层看，才能充分显示出这座塔的实际力量。从远处看时，电视塔显得头重脚轻，仿佛正在飞行。这座塔的原型是德国柏林亚历山大广场的电视塔。但布拉格的地理环境与柏林不同，

布拉格有更加蜿蜒曲折的山坡地景，所以尽管塔的尺度巨大，日什科夫电视塔仍然不能主宰这座城市的地平线。也或许是这样"天外来客"的造型，增强和突出了维特克夫山地的景观。

电视塔的修建跨越了共产主义和后共产主义时期，在电视塔建成之前，共产主义时代就已经结束。在拥抱未知的情境下，捷克著名艺术家大卫·切尔尼在2000年给电视塔上添加了10个戴着天线头爬行的巨型黑色婴儿。这个装置艺术作品在这个波希米亚式的诙谐和反传统表演中削弱了塔的竖向比例和象征意义，颠覆了这座塔的统治权威感。

今天，电视塔作为观光景点面向公众开放。

...rising, ambiguously, on the edge of the horizon.

It dominates, but with a surreal appearance of lightness. Despite its futuristic, technocratic and even alien appearance, it nonetheless resonates harmoniously in the landscape. It immediately became an icon of Prague — because it is mysterious while it is exposed. It stands at the perpetual threshold of being "other".

By 1974, there was a need for more radio and television signals over Prague (Perhaps also to interrupt signals from the West). The construction of the tower, by an industrial engineering company from Ostrava, took place between 1985 and 1991, with architect Václav Aulický and structural engineer Jiří Kozák.

It is the tallest tower in the country, and moreover it is built on top of a hill. Its antennae indicates the presence of the city from a great distance. A powerful and innovative steel construction, it produced many patents. Perhaps only at ground level, is it fully apparent how heavy an imposition the tower really is. When viewed from a distance, the tower appears top-heavy, as if taking flight. While it appears to float above the ground, in fact the opposite is true. To construct the tower and its terraced ground required the demolition of a city block and part of a Jewish cemetery.

Three double encasements of steel tubes stand upon a reinforced concrete foundation slab, 30m in diameter and 15m below ground. The 30cm thick space between the two steel tubes is filled with reinforced concrete. Within these three structural towers are the elevators and stairs. The main tube has a diameter of 6.4 metres while the others are 4.8 metres across. These structural tubes are 134 metres tall, and the tower reaches its maximum height of 216 metres weighing 11,800 tons.

The argument for 3 separated tubes was made so as to lighten its overall appearance and its perceived disruption on the horizon. It would be more structurally stable and the maintenance savings would also be less. These three towers support 3 separated glazed levels. The first is a restaurant 62 metres high, the second is a 93-metre-high viewing platform (with a 100 kilometre view), and the third is a service level at 130 metres.

This type of tower was proposed after the model from Berlin. However, the context of a tower in Prague, with its flowing hilly landscape is different. Despite its enormous scale, the Žižkov Television Tower still does not dominate the horizon. Despite its alien character, or perhaps because of it, this new civic mark enhances and highlights the landscape around Vítkov Hill.

Ironically, the Communist era in Prague had ended before the construction of the Žižkov Television Tower was complete. In 2000, in an embrace of the now unknown future, a project of artist David Černý added 10 crawling babies onto the facade. In a reformist and iconoclastic play, they undermine the tower's scale and symbolism. These alien babies, with their television heads, bring a human scale to the tower. They subvert the dominating authority of the tower and bring to it a touch of humour. (For a closer look, 2 of these bronze babies are installed at the Kampa Museum on Kampa Island). For added Bohemian humour, a hotel with a single room was incorporated into the tower in 2012, offering a view over Prague from its bathtub.

Today, there are exhibition galleries and a park at ground level. The restaurant and viewing platforms are open to the public, and at night, the Žižkov Television Tower lights up in the colours of the Czech flag.

当代

1993—

也许这个时期始于 1989 年天鹅绒革命和东西方铁幕落下，或始于 1993 年 1 月 1 日捷克和斯洛伐克的"天鹅绒分离"，或始于 2004 年，捷克共和国加入欧盟，或始于 2013 年，国家第一次总统直选，抑或始于 2018 年，第一个共和国成立 100 周年，国民对于国家成就的认同显得最为强烈……

"当代"一词在今天被广泛使用，但其边界很难界定。回顾历史，所有时期在当时也都被称为"当代"。准确地讲，对"当代"的定义离不开其历史语境。

在后共产主义时期，以国家为中心的社会结构瓦解，中央计划经济结束，自由市场经济重新崛起。人民、市场、国家和宗教的权威重新回归，建筑师也摆脱了国家控制的设计院所。随着思想解禁及经济自由化，国外投资迅速增长，布拉格的城市建设也开始复兴，许多长期废弃的地块又重新被填满，但出售的地产多用于商业而非文化和公共用途，包括"企业主巴洛克"风格的住宅都在此时出现。然而，这个新时期，也就是当代，继承了这个城市之前的传统。在用"萨拉米"方法开发布拉格的同时，人们开始重振这座城市的文化和遗产。

当时的总统瓦茨拉夫·哈维尔大力支持文化遗产保护，重视新建筑在重获自由的城市中发挥的综合作用。在过去和未来拉开帷幕的过程中，城市发展进入新的阶段，对其的公开讨论由此展开并持续至今。

20 世纪 90 年代初期，布拉格在某种程度上可以说是百废待兴。这座城市新的建设首先被国外著名建筑师（再次）引进，作为激发城市活力的推进剂。但是，在接下来的几十年中，随着捷克城市和社会的重建，更多的优秀本土及海外捷克裔建筑师登上舞台。他们遵循着捷克功能主义的传统，继续实践着所谓"捷克式"的严谨，简单的形体、优质的材料、令人印象深刻的细节，专注于精细的构图，把这种极简主义巧妙地融入历史环境中。

在布拉格密集的历史肌理中，这些优秀的当代建筑散落其中，它们更加尊重周围环境，共享着城市的文脉，代表着布拉格新的城市形象。通过跨越历史记忆与复杂历史环境的深入交流，每一幢新建筑都形成自己的信息、编码或主题，在城市复兴中共享相同的利益。

以下项目都在当代以不同的方式与布拉格的城市肌理相结合，且都可以被视为"神圣"的建筑，不是因为宗教原因，而是因为它们重新塑造了场所精神。

当代的布拉格
The Contemporary City

外立面印字
Text on Facades

玻璃面喷涂图像
Printed Images in Glass

曲形玻璃面
Curved Glass Facades

双层表皮
Double Facades

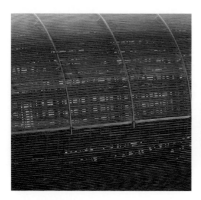

塑料和木表皮
Plastic and Wooden Skins

绿色屋面
Green Roofs

CONTEMPORARY

1993–

... perhaps already in 1989 with the Velvet Revolution and the fall of the Iron Curtain, or perhaps by January 1, 1993 with the "Velvet Divorce" of Czechoslovakia. It was perhaps in 2004, with integration into the European Union, or in 2013 with the first direct Presidential elections, or perhaps in 2018 with the 100th anniversary of the First Republic, that a search for fulfilment was its strongest...

The term "Contemporary" is, inevitably, widely used today. But its limits are difficult to define. Looking back through history, all times can be said to have been their own contemporaries. What can be said for certain is that the "Contemporary" is inseparable from its history.

The post-Communist time, in its sudden re-opening of old spaces, marked a moment of great euphoria throughout the land, with respect to local sovereignty, and with the re-integration of private markets into the daily system of the city.

The State-centric structures were decentralised and the authority of the people, the markets, the Republic and the church returned. Yet, the collapse of the authority of State-driven economics yielded other new forces as well: the resurgent primacy of the private markets. In terms of the city, this phenomenon is often referred to as the "salami method", whereby the whole is subdivided into small and separate parts. To some degree, the difference between hegemony and diversity was fused, as the all-encompassing development of the contemporary city produced the promise of "anything goes".

In the years after 1989 with the return of the market as the highest form of law, priority was given to private trade and merchants (investors and companies) over people and public culture. The city began selling its many properties for commercial uses, rarely for civic and cultural purposes.

The end of State planning and the ensuing liberalisation of the city rapidly initiated a new art of self-expression, private accumulation, and a vertical hierarchy within the social fabric. Again as in the 19th century, a variety of expressions and symbols appear as icons of individual identity, yet at this time, for the most part without the accompanying philosophical and cultural interest.

Nevertheless, the new (and perpetual) period — the Contemporary — inherited all that which had preceded it. Concurrently with the "salami method" of developing the city, came a renewed interest in revitalising the contexts, the heritage and the meanings of the city, along with its memories and ambitions.

The importance of culture and heritage and their synthesis in the new

35. 会跳舞的房子
Dancing House

36. 金色天使大厦
Golden Angel

37. 皇家花园的柑橘暖房
Orangery in the Royal

38. 鹿谷步道
Deer Moat Passage

39. 欧洲大厦
Palace Euro

40. 国家技术图书馆
National Technical Library

41. DRN/ 民族大厦
DRN/Národní Palace

42. DOX 当代艺术中心
DOX Centre for Contemporary Art

43. 布拉格艺术馆
Kunsthalle Praha

architecture of the free city was strongly supported by the new President Václav Havel. In re-opening this curtain between the past and the future, a new stage of the present was revealed and an open dialogue began which continues to this day, albeit quietly.

In the early 1990s the city of Prague was to some degree in ruins or falling apart. The new architecture of the city was to be imported (again) by celebrated architects from abroad. But, over the following decades as the Czech city and society re-established their collective autonomy, local architecture of a new kind began to appear again.

Inside their dense city fabrics, exceptional works of Contemporary architecture speak for the city and about a shared, historical substance. They are less self-referential and more responsive to their surroundings. Through a deeper communication across historical memory and within historically complex sites, each building with its own message, program or agenda, they share the interest in civic renewal.

The following Contemporary projects all merge with their landscapes in different ways and as such can be considered "sacred" architecture, not for their religious purposes, but for their re-grounding of the symbolic spirit of their place.

会跳舞的房子
Dancing House

地址：Jiráskovo náměstí 1981/6, 120 00, Praha 2, Nové Město
交通：Tram 1, 5, 14, 17 Stop Jiráskovo náměstí（2 分钟步行）
建造时间：1992—1994 年
建筑师：弗兰克·盖里（Frank Gehry）、弗拉多·米卢尼奇（Vlado Milunić）

■ 新古典主义/ NEO-CLASSICAL
□ 当代/ CONTEMPORARY

会跳舞的房子可能是布拉格最著名的现代建筑。这座位于伏尔塔瓦河畔的房子被构想成一个传奇,用来复兴民族精神,并成为一个新时代的标志。它的修建填补了一个双重空隙:在时间上横跨了共产主义和后共产主义时期;在空间上填补了一个被第二次世界大战炸毁并废弃的场地。这是天鹅绒革命和捷克与斯洛伐克分离后,捷克共和国第一个由外国建筑师设计的作品,代表着1989年之后捷克文化的复兴和对当代西方世界的回归。

这个项目毗邻瓦茨拉夫·哈维尔总统的旧居。20世纪90年代早期布拉格历史城区严重失修,1994年,在瓦茨拉夫·哈维尔总统的直接推动下,由荷兰保险公司投资,加拿大裔美籍建筑师弗兰克·盖里和捷克裔克罗地亚籍建筑师弗拉多·米卢尼奇合作实现了这个建筑,建筑师首次在设计中使用了3D计算机建模技术。

这是一种基于城市历史遗产环境的新设计。它利用了周边环境的特点来形成自身独特性,创造出一个叛逆的建筑图景,旨在将一个多重复杂的环境中激起多层次、多级别的城市对话。因此,建筑被要求形成一种对场地、街区、广场和河堤的"开放—闭合"形式。伏尔塔瓦河弯曲的弧线、城市公共建筑的纪念性、新艺术运动建筑为主的历史地块,

这些环境条件以一种复杂形式混合在一起。建筑用一种全新的语言呼应着与场所的联系,建筑自由流动的外观及建筑体上的波浪线条呼应着整个地块主导的新艺术运动建筑风格的浪漫曲线。建筑屋面的金属球网也是对维也纳分离派展览馆屋顶上镀金月桂叶片组成的花球的致敬。

建筑动感的形象好似一对舞者——弗雷德和金吉尔。通过建筑舞动的姿态去反映革命的精神和新时代的可能。建筑在静态和动态的力场中被建造,无论在城市还是社会中,它都形成了一种平衡的爆发。建筑两个主要的体块:一个透明玻璃的建筑通透体量和一个不透明的建筑实体互相结合,都面向布拉格城堡。建筑上面弯曲、分叉并且扭曲的柱子让人想起巴洛克建筑的那种繁复和雕塑般的张力。非线性几何图形将玻璃立面重新配置成具有雕塑般的表皮,两个建筑体量的轻快感都得到增加。玻璃幕墙后的地板仿佛飘浮在空中,而巨大的

建筑体积似乎在玻璃幕墙后面隐没。它像是回到了现代的透明自由空间——展示了流动的、数字化的时代。

立面上展现出的所有建筑视觉元素是自由的，它们在一个共同的运动中被结合在一起。交错的、突出的窗户和独立的结构件被放置在波浪线的背景上，这些线条与周围建筑的立面线条相呼应。凸出的阳台给建筑赋予一种人体尺度感，并增强了构图的拟人化特质——它像是一颗心脏，从建筑内部迸发，带着超现实主义的色彩。

与外部布局不同，建筑的内部是开放的平面空间，有着雕塑感和不规则的边缘。不同楼层的室内空间由不同的知名建筑师设计。顶层有酒吧和屋顶平台，站在屋顶的金属球形装置下可以远眺伏尔塔瓦河两岸的美景。

...along the river, in the midst of it all, a dance across the city.

The Dancing House was conceived to become a myth and to revive the national spirit. It was to fill a double-void: in time, spanning the Communist period, and in space, within a bombed-out and abandoned site. Part tower and part temple, the landmark was to be an icon for a new era: a contextualised innovation of heritage. It is of such significant uniqueness, that a nickname was necessary to express it: Fred and Ginger, the dancing pair. It pulls on its environment and forms itself within it, into an uneasily-defined, fully con-temporary image.

The project was also a form of union and experiment amongst its creators. In the context of severe disrepair in the historical city in the early 1990s, the Dancing House was realised in 1994, in cooperation with (then President) Václav Havel, and architects Frank Gehry and Vlado Milunić. Situated adjacent to Havel's former residence, one symbolic priority was communication with all of its neighbours. It exhibits a direct and explicit engagement with the distant Prague Castle, with the nearby tower of the Mánes Union, and with the houses at its sides. The project was given the right to build above city-owned land, and the city itself is a co-owner.

As the first symbol of a contemporary cultural renewal, the building could have been filled with many cultural functions. The building itself, however, is a cultural object (even while it functions as an office building). Through its dance it reflects the spirit of the revolution and the possibilities of a new time. Amongst static and dynamic fields of forces, in the city and in the society, it is formed and in turn forms a balanced burst.

The Dancing House establishes a civic dialogue on multiple levels and scales, within a complex, multi-scalar volume of civic space. In a return to the optics of perception, the building was designed to be seen and to be experienced in time. Its requirement to become a form of "open closure" to the site, to the city block, to the urban square and to the riverfront embankment — all at once — necessitated the first use of 3D computer modelling in design. Multiple iterations and volumetric models were used to prioritise a balanced building form, with respect to all its views and scales.

The bending curve of the Vltava and the monumentality of a civic tower mix together in this complex hybrid form. There are two main figures, both oriented towards the Castle, and they also combine with each other. A glass figure and a solid figure are lifted above the space of the ground. The bending, bifurcating and twisting columns recall the Baroque, in their doubling and sculptural tension. But they are without the classical motifs of a base and capital. The non-linear geometries and perceived tensions of the curved columns are transfused into the sculpted skin of the free-floating glass façade.

The lightness of the two volumes is enhanced with further play. The glass curtain wall reveals floor slabs which seem unsupported while the massive solid volume appears recessed behind its own protruding windows. Unlike most "postmodern" buildings, (as the Dancing House is often called) the structure here is exposed. In fact, it is more a return to the transparencies and free spaces of the Modernist time — although in a fluid and digital age. An additional layer of expression, not directly visible, is the employment of pre-fabricated concrete panels to produce irregularity and variety in contrast to the repetitive sameness for which it had been long known.

The visual ornaments on the facade express the idea that all elements are autonomous and free, held together in a shared movement. Staggered, projecting windows and individual structural clips are placed upon a background of waving lines which pin into the proportions of its neighbours' facades. The protruding balcony gives a human scale to the building, and enhances the composition's anthropomorphic qualities — with a touch of surrealism, it is a heart, bursting out from within.

The interior of the building is laid out in a different manner than the exterior. It is an open plan space with a carved and irregular edge. Different floor offices were designed by different famous architects. The top level exhibition gallery has a bar and a roof terrace. From beneath the perforated and bursting "New-Old" cupola on the roof, is a great view of the city centre.

金色天使大厦
Golden Angel

地址: Plzeňská 344/1, 120 00 Praha 5, Smíchov
交通: Tram 1, 5, 4, 7, 15, 9 **Stop** Anděl（1 分钟步行）. Metro B **Stop** Anděl（1 分钟步行）
建造时间: 1999—2000 年
建筑师: 让·努维尔（Jean Nouvel）

☐ 当代/ CONTEMPORARY

金色天使大厦被作为布拉格斯米霍夫老工业区更新的一部分，通过功能置换和创新设计，将现代建筑融入历史环境，用一系列新的功能：商业、办公和餐饮等，来复兴这个19世纪的老工业区。该项目由ING地产公司投资建造，法国建筑师让·努维尔1994年开始设计，与捷克斯帕莱蒂玻璃公司合作完成。它醒目的弧形玻璃体突显了捷克玻璃工业的卓越性，也证明了在一个具有创新性的历史城市中，新的市场开发的可能性。

金色天使大厦位于地铁B线Andel站，两条大街的交会处。这里是该地区一个重要的交通枢纽，交通位置优越，人流聚集，有多条城市及城际的公共交通线路经过。为了造就一个充满活力的公共空间，建筑在设计时尽可能地远离街道，以形成一个公共广场便于举办市集及各种活动。建筑屋顶有一个露台和花园，使之与地面广场呼应。建筑北立面轮廓线起始位置与其相邻的犹太教堂齐平，逐层升阶，向上升起，似乎从教堂的屋面直接连续上升到街角的顶点。新旧建筑之间通过这样的方式取得了联系。

在街角处，全玻璃的建筑似乎漂浮着，建筑与周围的城市景观融为一体并映射出周边的环境；同时，在建筑光滑柔和的玻璃曲面中，玻璃外立面也重塑了周围的环境，建筑上反射的波纹曲线也许是城市景观的隐喻。建筑外墙弧形玻璃上印有超过1亿个细点来表现布拉格城市的文学和图像，就像新舞台的玻璃砖外表面一样，这种模糊的幻影呈现出一种不可思议的虚幻力量，暗示建筑以未知的方式与过去及未来融合。大厦面朝历史城市，背靠新的商业区，通过深暗的颜色、流动的形式映射了老城的形态。在古老的城市环境中，金色天使大厦以新的姿态似乎在向前奔腾，表现出一种乐观的精神。

如果说会跳舞的房子是布拉格历史城市在新时代重新面向西方世界的象征，那么金色天使大厦则代表了城市老工业区面对历史和未来的态度。为了实现旧斯米霍夫地区的复兴，区域里的许多工厂和仓库都被改造。布拉格其他的老工业区，如卡林和霍拉舍维采，也正在以各种方式对这些工业遗存进行保护与再利用。

… on the edge of the historical city, a new structure pulls out of the old, engages with it, and at the same time disappears.

Printed on its rounded glass facade, is a phantom image, mysteriously hinting at different and unknown ways to integrate heritage and future. Within the fluid fabric of the city, the Golden Angel appears to be surging forward. It faces the historical city and behind it stands a new business district. As the Estates Theatre echoed sound and silence simultaneously, here there is a projection of multiple shades of light back onto the city, as the building also reflects the old city in its dark, fluid form.

Optimistic in spirit, its form weaves into the fabric of the city by offering an urban opening (the plaza on its eastern side) for the free movement of people. As the Dancing House is a symbol of the entire city in its new era, the Golden Angel represents a renovated neighbourhood and its local environment. Named after the demolished statue of a golden angel, even the building's name symbolises a transformed renewal.

It is part of a much larger project, and was not planned to stand alone as an infill building. The construction of an entire modern neighbourhood, called New Smíchov, was to replace the industrial district of the 19th century with a range of programs, shops, offices and restaurants. Rather than invest in returning the past into the present, the decision was to forge forward with something new, in an attempt to integrate modern offices buildings into the historical environment. The project was built for ING Real Esate in 1999 by Jean Nouvel, in cooperation with the Czech glass company Sipral. It highlighted the excellence of the long-standing Czech glass industry. It also proved that the new market development was possible within the context of an innovating, historical city.

Linked into a main intersection, along the north-south axis leading north towards the Prague Castle, the Golden Angel is the representative tip of the renewed neighbourhood. By placing void spaces into the composition of the building, the space of the city is mixed with the built structure. New types of airy, porous office buildings which are delicate and light integrate into the surrounding city landscape.

The building recedes from the street as much as possible, in order to create a public plaza, to let the city itself come to life beneath it. At the corner, the tower appears to be floating overhead. Its double-height arcade spans the full length of the block. The upper level offices set back even further again from the arcade, offering on the 3rd level, a terrace and a roof garden, tying the roofscape back to the public plaza on the ground.

A stepped terraced form appears to rise up from the roofline of the adjacent synagogue towards its climax at the inset corner. It appears as a series of levels ascending from the past, with the lines of its facade connecting to adjacent buildings. The smooth flow from old to new accentuates the tower's verticality and at the same time reduces the perception of its overall mass.

The fully glazed facade reflects its surroundings. But in sensual and smooth, gentle curves, the curtain wall also reshapes them. The ripples and curves reflected in the tower are perhaps a metaphor of the landscape of the city. The mysterious figure of a subtle spatial depth is printed on the glass of the tower with over 100 million dots, mixed with texts and images of the city. Like the sliced glass skin of the New Stage, this ambiguous apparition presents an uncanny illusory force.

In order to realise this new district, many of the manufacturing and storage structures in "old" Smíchov were demolished. The industrial characters of other more recently redeveloped neighbourhoods such as Karlín and Holešovice, on the other hand are being reintegrated into the developing city in a different manner.

皇家花园的柑橘暖房
Orangery in the Royal Gardens

地址: Královská obora, 118 00 Praha 1, Letná
交通: Tram 22, 23 Stop Pražský hrad (4 分钟步行)
建造时间: 1999—2001 年
建筑师: 埃娃·伊日奇娜 (Eva Jiřičná)

■ 文艺复兴／RENAISSANCE
□ 当代／CONTEMPORARY

皇家花园的柑橘暖房与安娜王后的夏宫一步之遥，建在原文艺复兴时期温室的遗址上。这是一个在历史环境中使用全新材料和结构的建筑，通过鲜明对比来反衬历史和传统的同时，又通过先进技术来展现当代，实现复杂环境中的微妙平衡。

柑橘暖房最早建于 1560 年，一直以来为布拉格城堡种植和提供植物花卉，但在瑞典军队入侵的三十年战争中被毁，之后数百年一直是废墟。20 世纪 50 年代温室进行了初步修缮，直到 20 世纪末，重建工作才彻底完成。该项目是瓦茨拉夫·哈维尔总统为了纪念其第一任妻子奥尔加·哈弗洛娃而发起，委托捷克裔英国建筑师埃娃·伊日奇娜设计，项目于 2001 年建造完成。

这是建在斜坡上的管状全玻璃建筑，巨大的空腔结构由不锈钢管组成拱形截面并相互连接成一个整体。钢化玻璃通过支点连接在这些支撑上，屋面设有可以开启的三角形天窗，在没有任何室内立柱的情况下实现了 84.5 米长，9 米宽和 5.2 米高的全玻璃外壳建造。建筑分为三个不同功能的区域，室内空间具有灵活性并可以适应种植各种植物。通过计算机和遥控装置可调节温室内的热量、通风、温度和照明，先进的技术使得温室内四季如春，建筑适应性得到进一步加强。

尽管使用当代材料建造，柑橘暖房也在许多方面呼应了其周边环境。建筑精湛的结构技术以其清晰的构造表达，映射了圣维特大教堂的哥特建筑特征，也彰显了捷克当代的玻璃技艺。通过与其南部边缘现有的道路连接，它又被巧妙地置于周边的景观当中。建筑北墙又建在文艺复兴时期的建筑基础上，从它的两侧看，温室似乎是从大地中生长出来。

皇家花园的柑橘暖房，有着它自己独特承载光和土地的形式，孕育了新的生命。它可以被看作是在大地和太阳交会处的瓦茨拉夫国王陵墓对位的女性形象。这是一个承载永恒生命力的空间。年复一年，四季不断，鲜花盛开。

… at the heart of the Royal Gardens, to the north of the Prague Castle, where it has always been.

For over 450 years, this site has served a single purpose — to grow plants and flowers for the representative events taking place at the Castle. The original Orangery was built in 1560, but was destroyed in the 30 Years' War by the invading Swedish army, and for centuries remained a ruin. Preliminary renovations from the 1950s were technically insufficient, and by the end of the 20th century, were finally replaced. The contemporary Orangery is built upon its original Renaissance foundations.

It is a small scale renovation within a fine and detailed fabric of heritage. Only a step away from Queen Anne's Summer Palace, it is not intended to be representative of privilege, reserved only for private summer use. Rather, it is a component of a larger complex, functioning year-round as a service building. It is connected to the maintenance of the Castle and the garden, and as a contemporary greenhouse, makes use of advanced technology to be functional throughout all the seasons.

The renovation was commissioned by President Václav Havel. It was realised (in remembrance of his wife Olga Havlová) between 1999 and 2001 by the Czech-British architect Eva Jiřičná. It is an interpretation of the Renaissance winter garden with a new material palette. As the project symbolised the revitalisation of historical memory and tradition, it also demonstrated the novelties of the contemporary period through its use of advanced technologies. It is not "new" for its own sake, but to achieve the delicate balance of its complex environmental requirements.

The semi-circular, glazed construction is nearly 100m long. It is a massive structural tube — a curved facade that is both wall and roof — that is supported by exterior stainless steel construction. The building is divided in 3 parts, with 4 semicircular arch supports. Tinted tempered glass spans between these main supports, in a cross-braced web of welded connections. This structural solution allows for the 84.5-metre-long, 9-metre-deep and 5.2-metre-high fully glazed enclosure to be built without any interior columns. The interior is therefore fully flexible and can accommodate a wide range of plants. The adaptability is furthered through the use of remote controlled shading, ventilation and lighting.

Despite its contemporary material palette, the Orangery complements its historical environment in many ways. The highly-engineered external structure, with its expressively articulated parts, echoes the Gothic character of St. Vitus Cathedral. It is delicately placed into its landscape along the existing path on its southern edge. At its sides, it appears to emerge from the earth, and it is built into the Renaissance foundation wall on the north.

The building is its own form of temple — it is a house for the life between light and land. The Orangery in the Royal Gardens, physically and symbolically captures the sun above the Castle and transforms it, through the earth, into a new form of life. It can be considered as the female counterpoint to the tomb of King Wenceslas at the crossing of the sun and the land — as the space within its own sphere of perpetual life — growing flowers year round.

鹿谷步道
Deer Moat Passage

地址：U Prašného náměstí, 118 00 Praha 1, Hradčany
交通：Tram 22, 41 **Stop** Pražský hrad（5 分钟步行）
建造时间：1999—2002 年
建筑师：约瑟夫 · 普莱斯科特（Josef Pleskot）

■ 文艺复兴/ RENAISSANCE
□ 当代/ CONTEMPORARY

鹿谷步道是布拉格城堡北面的雄鹿城壕中一条连接城堡的人行步道。该项目由瓦茨拉夫·哈维尔总统代表国家委托约瑟夫·普莱斯科特设计，历经3年建成。这条步道以人行的尺度，简单的材料穿越了这里的自然地景并融入山谷，以一种新的形式连接了这片古老的土地，成为一个具有象征意义的建筑景观地标。

鹿谷步道沿着城壕的山坡而建，通道为木板搭设，设金属护栏，宽仅适合一个人的通行，这样的尺度设计是为了增加来往游客相遇侧身通行时产生交流的可能性。从远处看，通道简单的形体及构造几乎无痕地融入环境。这段步道设计的独特之处是穿过布拉格城堡北桥下面的隧道。隧道沿着雄鹿城壕中四季流淌的溪流的自然路径而造，全长近84米，宽1.5米，其中有78余米埋入山体。隧道入口的挡土墙在起始位置由毛石砌筑，这种材质与铺地匹配。之后墙体变成了浇筑混凝土，并在表面保留了木制模板纹理。混凝土入口从山体里伸出，洞口里散发出微弱的柔光。

隧道内的地坪被分成两部分材质：一部分用与天然石材相似的预制混凝土砖，具有细密条状肌理；另一部分用连续的金属格栅盖板，格栅盖板下是流动的溪水，水声在整个空间轻声回响。隧道是一个连续椭圆拱结构，用深红色的砖砌成，让人同时联想到罗马、巴洛克和现代风格。每块砖之间的填缝都有轻微的凹进，使每一块砖都具有立体感。砖块按照交错模式竖向砌筑，在表面产生了微妙的韵律变化。安装在地面内的嵌入式照明在隧道的内表面投射出的弯曲光弧也增强了这种视觉效果。尽管隧道很长，但是它小尺度的材质还有整体空间构成比例却给人带来一种亲切感。隧道中的每一块砖，砖块之间的组合以及其他建筑细节都和谐统一地构成了这个空间，让人体验深刻。

如今，它在春夏季节短暂向公众开放。

… in the valley, a pathway flows along a stream — naturally — up to Prague Castle's northern gate.

It is a literal and a symbolic landmark reconnecting with the ancient land. Sensitive and delicate, a detailed and finely crafted element, it is woven into the valley. Commissioned by President Václav Havel on behalf of the country, it was part of an initiative to bring a new path through the Castle grounds. It is part of a larger sequence of pathways from the northern slopes of the Castle, through the Deer Valley, upwards towards the northern gate. It serves as an alternative way into Prague Castle and follows the natural path of the stream.

Where waters once drained to forge the landscape's form, and where the Brusnice stream still flows, this new pedestrian footpath is fully integrated into its natural environment. It was built by architect Josef Pleskot between 1999 and 2002. The Deer Moat Passage cuts through the massive earth berm beneath the Renaissance-era bridge overhead, to make a path which maintains a human scale. As a natural piece of the ancient landscape and within the protection of the forest and the Castle, it belongs to the people.

Its main element is the tunnel. It is 84 metres long and 1.5 metres wide, and travels 78 metres underground. As one approaches, the natural landscape gradually becomes a built structure. The retaining walls which frame the tunnel openings emerge as rough stone and match the stone paving. They become walls of cast concrete (which are treated so as not to age), and they retain the wooden formwork within their textured surface. These walls converge at the point of entry and define an iconic portal leading into the earth. This exposed concrete portal projects out from the land and reveals a soft illuminated glow within.

The ground within the tunnel is split into two halves. On the one side are prefabricated concrete tiles, textured with a natural stone relief. On the other side, a continuous steel grate covers the flowing stream which runs along the length of this path. The water is exposed beneath, within a half pipe of concrete, and its sound softly echoes throughout the interior.

The tunnel is constructed as a continuous free-standing arched construction within a larger, unseen, concrete vault. The shell which defines the interior space is built of dark red bricks, in a particular, elliptical form that is a fusion of the Baroque, the Romanesque and the Modern. Mid-way, the exposed foundations of the original pier of the Renaissance bridge emerge, as if an ancient temple revealed. This masonry pier is finished in a cement stucco, which is treated with steel to match the colours of the bricks.

The architectural composition is unified by the expression of each brick, the criss-crossed pattern of their layering, and the perfectly aligned details of all the edges. The mortar between each brick is slightly recessed, which enhances the perception of each individual brick. Their assembly, moreover, is placed along a staggered pattern, to create a subtle image of reverberation. The angled walls of the tunnel smoothly become its ceiling. This visual quality is enhanced by the recessed lighting which is placed within the floor to project curved arcs of light along the tunnel's surfaces. Despite its great length, through the texture of its materiality and the scale of its composition, the tunnel has a feeling of intimacy.

It is open to the public in the spring and summer seasons.

欧洲大厦
Palace Euro

地址：Václavské náměstí 772/2, 110 00, Praha 1, Nové Město
交通：Metro A, B Stop Můstek（1 分钟步行）
建造时间：2002—2004 年
建筑师：DAM 建筑事务所（DAM Architects）

■ 哥特/ GOTHIC
■ 现代主义/ MODERNISM
□ 当代/ CONTEMPORARY

欧洲大厦是天鹅绒革命后在布拉格历史城区建设的第一座全玻璃外观建筑。大厦坐落在老城和新城交界处：穆萨克（捷克语：小桥），护城河街和瓦茨拉夫广场的交会点上，这里也是从老城进入瓦茨拉夫广场的门户。

基地上原建于18世纪及19世纪的房屋在20世纪70年代的地铁建设中被摧毁。在之后的几十年里，专家为此地的建设提出许多设想，并意图发展它作为城市的文化或信息中心，使其在城市结构中可以实现公共利益。21世纪初，该块土地被私人购买，业主希望在市中心建设办公楼来代表市场经济的重新回归。由于这一位置的特殊性和重要性，在此基地建设的讨论中，一场关于城市遗产保护与设计创新表达、市场私有利益与遗产公共利益之间的博弈就此展开。最后，双方互相妥协并达成了平衡。

DAM建筑事务所在2000年赢得这个场地"欧洲大厦"的设计竞赛，在经过两年的协商后才开始建造。在面对连接两个广场、周边不同时代、不同风格建筑的复杂环境时，采取了全新的方式与历史环境进行对话。建筑地上十层，地下三层，功能主要用于商业及办公。空气动力学形状的连续全玻璃外立面包裹着这个三角形的基地，削弱了建筑的体量感，并反射出周围的历史环境。从每一个面，它都推动了建筑与周边进行一场特定地点的特殊交流。玻璃外墙与相邻建筑基本同等高度，保持了体量的一致性。玻璃外表皮中插入一个金色的立方体塔楼，由可调节的金色百叶围合，以满足室内照明的舒适性。双层表皮之间用于通风，使得建筑节能环保。金色的立方体伸出玻璃体，显示出一种高贵而现代的质感，并与对面科鲁纳宫（建于1914年）的新艺术运动风格的塔顶新旧对应，一起成为瓦茨拉夫广场的左右门户，并变成广场重要的节点。

如今，欧洲大厦和科鲁纳宫，这两个建成历史相差近百年的建筑，具象征意义地一同对称矗立于瓦茨拉夫广场的入口处，这不禁引出了一个问题：如果这两个建筑只存在一个，这座城市看上去又会是什么样子？

… at the epicentre of the city, a new civic tower weaves into the past.

At the intersection of the Old and New Towns, at Můstek — the meeting point of Na Příkopě Street and Wenceslas Square — stands a contemporary office building. Můstek is a unique moment within the structure of the city. This space is not a square, but rather an intersection of two other squares which are actually boulevards. It is the waiting space before entering the Old Town, and also the location of a Romanesque bridge into the medieval city. This small bridge (Můstek in Czech) was discovered during the construction of the metro in the 1970s.

Due to the significance of this site, in the construction of the Palace Euro, a struggle played out between the limits of urban heritage and the freedom of a new expression. The political and ideological debate — between the private interests of the market and the public interests of heritage — was absorbed within the building's form. The question concerned the degree of emphasis either on the private tower, or on the contextual resolution of the urban block. In the end a certain balance was achieved between the two competing interests, as each made compromises towards a formal synthesis.

The 18th and 19th century houses which stood on this site were destroyed during the construction of the metro in the 1970s. During the following decades, many proposals were made for this site, representing its potential as a cultural or information centre for the city, and its public purpose within the structure of the city. In the early 21st century, however, the land was purchased by a private owner, with the wish to place an office building in the city centre. After a competition and years of negotiations, the Palace Euro was constructed in 2004 by DAM architects.

The building is woven into its compact historical surroundings. Despite its 10 stories in height, its curved, recessed and accented massing fits into its various adjacent environments. The Palace Euro communicates with its context on many levels and from multiple points of view. The building is seen from three different sides, and from each it advances a particular and site-specific dialogue with its neighbours. The continuity around the building is achieved with a smooth wrapping of its form, which also diminishes its dominance.

The northern facade is subdivided into two visually distinct parts, to maintain the proportions established by neighbouring buildings. On the western corner, the slight overhanging volume overhead adapts to the curved form of the narrow historical street and reinforces its Gothic scale. From the south, the glass facade separates from the structure behind it, and becomes a free-standing wall that aligns with the curved roof of its neighbour, Ludvík Kysela's Pálac Astra.

Wrapped within its glass envelope (which reflects the site and context) is a golden tower. This internal tower is only seen from certain views. It is defined by a series of vertical brass fins and rises up from within the building mass. The tower exceeds its glazed envelope at the roof and projects a golden cube. This element which emerges at the top of the tower marks a symbolic gateway to the Old Town, and an echo of the Palace Koruna on the other side of the square. Together, these two towers as a pair frame this critical point in the city at the base of Wenceslas Square.

Today, the symbolism of the two palaces, like the two currencies, the Euro and the Koruna, standing together at the epicentre of the city, brings up the question: how would the city look with only one of them?

国家技术图书馆
National Technical Library

地址：Technická 2710/6, 160 00 Praha 6, Dejvice
交通：Tram 8, 18 **Stop** Lotyšská（4 分钟步行）
建造时间：2006—2009 年
建筑师：Projektil 建筑事务所 (Projektil Architects)

■ 现代主义/ MODERNISM
□ 当代/ CONTEMPORARY

国家技术图书馆位于远离市中心的捷克理工大学校园内。建造一座新技术图书馆的想法始于20世纪90年代，捷克教育部在2000年发起了一场设计竞赛，在近五十个方案中，Projektil建筑事务所的设计脱颖而出，获得了一等奖，并在2006—2009年间进行了该方案的建设。

图书馆布置在一个在75米×75米的场地上，建筑地下三层，地上六层。书库、技术设备、基础设施、供应与储藏，以及占据一层楼面的停车场都分布在地下。在地面层和第二层设主入口会议大厅、展厅、餐厅、书店和衣帽间等服务空间，所有这些功能都围绕入口大厅排布，入口大厅也是图书馆内部乃至整座大学校园的主要会面地点。建筑在每个主要方向上都设置了入口，对其周边展现出一种平等的氛围。建筑上层是图书馆空间，布置有书架、研究室、办公室和自习教室等。建筑设计的构思是不仅要为大学提供一座图书馆，也要使其成为这个大学的新的核心和活力中心，以及人们知识学习和交流的源泉。

虽然图书馆从外部看起来是纯净而封闭的玻璃几何形体，但建筑内部却是形态丰富、多层次的复合空间。内部中庭自然采光，并连接6个层面，每层地面用色大胆。进入图书馆后，人们将会惊讶且兴奋地发现一个缤纷的垂直空间展露在他们面前。中庭的混凝土墙面上绘有两百多幅漫画，它们是罗马尼亚先锋派漫画艺术家丹·皮诺舍维奇的作品。

建筑的室内空间极为开阔，混凝土的结构体系由跨度15米的双向网格组成，可实现室内布局的灵活性和可变性。可移动定制家具的设计给读者带来自由感。学生能够移动和重组家具，以形成自己的组合与座位排放方式，突出了最初的室内设计原则：合作与互惠。主要的视觉要素是用色大胆的地面，彩色的地板如同热力分布图，象征着空间中的能量分布和其流动。多姿多彩的室内表现使该建筑变为一个欢愉和动感的活力空间。

建筑的外玻璃幕墙是一种能控制热量吸收并进行自然通风的双层表皮。白天可以反射出周围环境景象，并透过玻璃显示出均匀间隔的楼板。而在晚上，图书馆内部的照明变成了立面上新的装饰。这种昼夜间的动态变化进一步强化了该建筑的反纪念性，为城市空间带来了惊喜。国家技术图书馆使人与知识在一个新的"自由和流动的体量"中融合，并具有了象征意义。

… at the off-centred centre of the off-centred campus.

As a library, it must represent itself as an institution, or at least extend its history. As part of the unfinished campus of the Czech Technical University, north of the Prague Castle in Dejvice, it is also a unification symbol. Despite its monumental form, it is a non-imposing structure. It is smaller than the other buildings of the campus, and does not dominate. It is a form of counter-monument.

The unfinished masterplan of the university dates from the 1920s. It was built in phases over the following decades, which produced a fragmented whole. A competition initiated by the Ministry of Education in 2000 resulted in the winning competition being constructed between 2006 and 2009 by Projektil Architects.

There was a lack of a clear central point within the urban campus. The National Technical Library was thus to become a symbol and a new centre of the University. While the site of the building is not in the centre of the area, the Library nonetheless acts as a social centre. It is a contemporary temple — without a plinth — and becomes a source of public knowledge and interaction belonging to the school and to the people.

Its rounded square form, of 75m×75m, is wrapped at ground level with clear glass, which makes the upper volume wrapped in vertical glass fins, appear to float. Its open ground level plan is accessible for the public, and reinforces the non-hierarchical concept of the public grounds. The object can be approached from all directions, articulating the equality of all that which surrounds it. The library's 360 degree unity holds together the diverse collection of its adjacent, disconnected buildings.

While it appears closed from the outside, within the library is a single interconnected interior. Upon entering the library from below, a colourful, vertical space is revealed in a surprising and uplifting moment. The internal atrium links 6 floors of different functions, each marked by different colours on the ceiling and the floors.

The first and second levels are open to the public. The spaces for social mixing, such as the café, exhibitions, and meeting places for visitors and students are located here. On the upper levels are the library spaces, study areas, administration, and classrooms. The naturally lit atrium organises and centralises the interior around a common space of light.

The openness of the interior operates horizontally as well as vertically. The 15m×15m internal grid allows for the flexibility and variability of the interior layout. The free use of space is defined by a range of seating choices, which encourages overlaps and mixtures of groups. The coloured floors represent this dynamic as a topographical energy map. It encourages the free flow of movement throughout, while indicating the hidden forces within the building. Its colourful expression makes the institution into a playful symbol, characterised by light, colour and the open plan.

The Library's glazed facade presents another dialogue between the exterior and the interior. Its appearance changes from day to night. During the day, it reflects its surroundings, and expresses evenly spaced floor slabs. However, at night, the various double height spaces of the interior are revealed. The illumination of the Library from within becomes the new ornamentation of spatial play filtered through the translucent facade. This dynamic between day and night further enhances the counter-monument quality of the institution by transforming its own image. It also brings a dialogue of diversity and surprise into the city space. The National Technical Library enables and symbolises the mixing of people and knowledge while advancing the Modernist "free plan" into a Contemporary "free and fluid volume".

DRN/ 民族大厦
DRN / Národní Palace

地址：Národní 135/14, 110 00, Praha1, Nové Město
交通：Tram 1, 2, 9, 17, 18, 22, 25, 41, 93, 97, 98, 99 Stop Národní divadlo
　　　(2 分钟步行)
建造时间：2012—2017 年
建筑师：菲亚拉和涅梅茨建筑事务所（Fiala+Němec Architects）

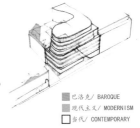

■ 巴洛克/ BAROQUE
■ 现代主义/ MODERNISM
□ 当代/ CONTEMPORARY

民族大厦位于民族大街的文化地带上,处在国家大剧院和 Máj 百货公司之间。这里是布拉格老城最后一块"空白"地,很长时间都处于废弃状态。1989年11月17日,5万名学生在这里集会引发了天鹅绒革命。因此,地块在城市中的特殊性要求其新建筑要成为具有象征意义的新地标。

这是一座新捷克风格的办公建筑,民族大厦在其新旧关系的辩证中颂扬了场地的历史和文化特征,并为集体记忆带来新生。整个建筑在各种捷克历史意象的相互交织中被赋予了民族精神。它就像对面料和碎布进行重修整合,编织出一副包含着历史记忆和当代精神的全新挂毯,并以独特的方式突出了建筑的自主、多样、可变、交互和精致的特征。

民族大厦试图以新的方式与周围环境融合,并与城市结构产生联系。从民族大街上看,大厦位于一个洛可可式住宅和一个现代风格的办公大楼之间,并且各个建筑都有着不同的高度和退界。为了将周边建筑和历史环境整合在一起,建筑立面被塑造成起伏的形状,并通过逐层退台来减弱建筑的高度和体量感。新建筑被包裹一个在深色、透明、光滑的玻璃外立面中,新型三层玻璃幕墙在太阳能控制、透光和隔热之间达到了平衡,使建筑能始终保持舒适感。如柳枝般婀娜的金属细条密集地附着在外立面上成为建筑的外表皮,呼应着这座城市新艺术运动风格的历史特征。一部分老建筑被整合到新建筑中,也限定出一个容纳新旧的院落空间。无论从时间、概念还是形体上,这座建筑从里到外都是混合的。

传统和历史在这座新建筑中交织。各种材料及建筑要素在时空中交融在一起,来自历史建筑的结构及材料在新建筑中被重新组装在一起。裸露的钢板梁与老的石梁相互穿插,方形的巴洛克木地板与混凝土砖混合在一起作为地面材料使用,不分彼此。建筑的天台可以说是一个公园,俯瞰着这座城市天际线。起伏在外立面上的各层阳台种满了鲜花:红色、蓝色和白色——这是捷克国旗的颜色。

作为一个重要历史街区的节点,民族大厦要反映出波希米亚的文化、品质和特征。因此,该建筑昂贵的造价使得建筑中的手工和定制产品细节都达到了与市民会馆相当的精美程度。对于办公建筑而言,这样的做法是不常见的。

... in the last "empty" parcel in the Old Town, space is already filled with memory.

"A Czech building". In an interwoven play of old and new fabrics, the Národní Palace celebrates the history of its site, its symbolic cultural renewal, and it brings a new life to collective memory. The building frames a courtyard as a space which is both old and new. This complex as a whole gives a form to the National spirit in an intertextual weaving of Czech iconography. It is a fusion of fabrics and fragments into a new, old-new tapestry. The flexibility, spontaneity, diversity, interaction and refined exchange of elements and traditions are highlighted in this building in a unique way.

The exceptional position of the building's situation within the city demanded a symbolically representative expression. This was the site where 50 000 demonstrators were brutally stopped on November 17, 1989 at a rally which became the Velvet Revolution. Something special had to be built on this piece of land. It is located between the National Theatre and the Máj Department Store. Despite its long history as a vacant site, it was full of memory and meaning. Between 2012 and 2017, the Národní Palace was realised by architects Fiala+Němec, with close cooperation between the builder and the client with the collective intention to create a symbol.

The Národní Palace blends and bends to its neighbouring buildings. On Národní, it is located between a Rococo house and a Modernist administration building, each of different heights and setbacks from the street. Its curving form and facade are shaped to join together these different edges and histories. The new building is situated within the imaginary, smooth envelope implied by the difference between the contextual edges of its site. Within these outer limits (the constraints of its history), its form is free.

Not only conceptually, but physically, the buildings are mixed. Their fusion creates a double-hybrid of new-old and old-new. There was a mural on the blank wall of the site, for many years, which characterised the prior period of time (or perhaps all time): an infinite loop of tanks and bulldozers. The spirit of this expression — a play of destruction and reconstruction — was reenacted through the making of this building. A piece of the demolished mural itself was re-integrated into the office building in a blurring of its past and present, its making and re-making. Part of its contemporary composition includes an existing Baroque house, which is to be further expanded into a new complex of spaces for UMPRUM, the Academy of Arts, Architecture and Design.

Within the building, tradition and heritage are literally woven with the modern and the new. Disassembled materials from the historical building are physically re-assembled within the new building. Exposed steel bracing is interspersed with those of original stone. Squares of Baroque-era wooden flooring are mixed with cast-concrete tiles. Repurposing and reintegrating collective memory is different from historicism. Materials are communicating with each other across time and space. The physical reality of the past is blended with that of the present, in a dialectical play and fusion of inversions.

Throughout the Národní Palace are signs of life. On the rooftop is a park; a public space upon the skyline overlooking the city. The undulating balconies are planted with flowers — of red, blue and white — in the colours of the Czech flag. Within the courtyard is an imaginary tree built of coloured metal, plants and glass.

This high degree of laborious detailing, of arts and crafts, is unusual for an office building (and it is one of the most expensive office buildings in Prague). But the builders' priority was to represent the national spirit, the experimental and free-spirited yet sophisticated building culture and the ironic humour of Bohemia. The multifold textural expressions and surprises found throughout the building not only symbolise — but a manifest — the freed limits of a Czech independence.

DOX 当代艺术中心
DOX Centre for Contemporary Art

地址：Poupětova 1, 170 00 Praha 7, Holešovice
交通：Metro C **Stop** Nádraží Holešovice (10 分钟步行)
建造时间：2003—2018 年
建筑师：伊万·克劳帕（Ivan Kroupa）、马丁·赖尼什（Martin Rajniš）、
　　　　彼得·哈耶克（Petr Hájek）

■ 新古典主义/ NEO-CLASSICAL
□ 当代/ CONTEMPORARY

布拉格 7 区霍拉舍维采，是一个始建于 1888 年的滨水老工业区。这个地区在 20 世纪后期处于极度衰败的状态。近些年来这个区域开始发展复兴。DOX 当代艺术中心就建在这里一个老钢构件装配厂之上，现在 DOX 已成为这个正在更新的老工业区的新地标。

DOX 是布拉格第一个展出当代艺术的私人机构，一个容纳专业学术、艺术创作和演讲交流等多重文化功能的开放综合体。DOX 的含义来源于古希腊语，意思是"感知的方式"，其寓意并回应这个工业区的复兴。2002 年，莱昂·沃卡购买这块地产并兴建了这个艺术中心。目前，艺术中心包含有三个主要部分：DOX 一期、格列佛和 DOX+，整个项目的规模还在持续扩张中。作为一个建筑作品，DOX 从内部更新城市结构；作为一种文化产品，它通过对意识形态差异的消解，使当代共同话语的讨论成为可能。

DOX 一期由建筑师伊万·克劳帕在 2003—2008 年之间完成，包含展厅、咖啡馆、办公室、会议室和图书馆等，其中庭院也可成为展示空间。DOX 一期的目标不是"最大化"开发，而是通过最少地加建、填充和扩建现有的建筑，去融合新旧建筑。它的内部结构就像一个由重叠和相互联系的空间组成的迷宫，内部空间由滑动的钢板进行分割，可以自由开合。

DOX 的二期格列佛，由建筑师马丁·赖尼什在 2014—2016 年之间实现。这个 42 米长的"齐柏林式飞船"由钢木构建而成。它的形状像一个飞行目的未知的飞艇，悬浮在庭院的空中，可能即将着陆、坠毁、升起或漂浮。其灵感来源于《格列佛游记》，代表了建筑师和艺术中心一种乌托邦式的理想。飞艇内部是一个小舞台，可以作为一个艺术公共空间来使用。作为探索未来以及场所复兴的召唤，无论是过去还是现在，内部还是外部，格列佛都是一个当代装置艺术的典范，是隐藏在城市街区的超现实主义的象征。通过对集体想象的扩展、再利用、再生产，激活了唤起新事物、旧事物以及其他各类相关事物的可能性。

DOX 的第三期 DOX+，由建筑师彼得·哈耶克在 2016—2018 年之间建造完成，其规模延伸到邻近的建筑和庭院，包含多功能演艺厅、排练厅、咖啡馆、教学及办公空间等。新设计的演艺厅和排练厅在扩建的庭院里被建造为独立的建筑体量，演艺厅可容纳 200 多人观演，并将 DOX 一期和 DOX+ 在内部进行连通。演艺厅内部与 DOX 一期的内部空间白色装修不同，它采用了灰色的素混

凝土。建筑外表皮采用灰色聚合物薄片构成的软垫进行包裹，显得十分柔软，与坚实的内部空间形成了鲜明的对比。

就像它所处在的更大街区乃至包围着这一街区的更大区域一样，DOX 艺术中心可以被认为是一种独特的岛屿结构类型。它处于不断成长、变化、适应、消化的动态过程中，在时间的长河里，DOX 艺术中心将区域的开放性、边界以其自身的形式，作为一种象征保存下来。

…in a renovating industrial district, a new centre is at once city block and landscape … and something else.

The meaning of DOX comes from the ancient Greek Doxa, "ways of perceiving". It also recalls the industrial heritage of the riverfront district and its former docks. This first private institution for contemporary arts, and for the experimental space between, has been expanding over 10 years. As the building is an assembly of multiple structures and ideas, so are the contents of its displays. It represents and houses an open synthesis of disciplines, artists and discourse.

Founded by Leoš Válka, who purchased the property in 2002, DOX is presently defined by three main parts: DOX I, Gulliver, and DOX+. With an aim to promote contemporary arts, the project intended to blur disciplines, to fuse past and present, and to bring together different people and actions around common themes. As an architectural artefact, it has renewed the abandoned industrial block from within.

Like its larger block and like the larger district which surrounds it — the complex of buildings at DOX can be perceived as a particular type of interwoven island. Prague 7, Holešovice, is an emerging modern counterpoint to the historical city. Established as a district of industrial enclaves in 1888, its identity is defined by self-contained campuses for a range of infrastructural needs: a slaughterhouse, energy production, a brewery, manufacturing etc. In the late 20th century, the area was in extreme disrepair. It did not experience a burst of renewal of development until relatively recently. DOX was built upon and within the existing, historical industrial buildings of a steel assembly factory. In 2008, after 5 years of work by architect Ivan Kroupa, DOX I was opened to the public.

DOX I did not aim to "maximise" development. Through minimal additions, infilling, building up and widening existing buildings, it blended old and new in a unified colour palette of white and grey. Its internal organisation is like a labyrinth of overlapping and interconnected spaces. Its temporary partitions are controlled with sliding steel walls which give an option of opening and closing both spaces and functions. There are multiple galleries, presentation halls, a café, offices, library and meeting rooms, all with natural lighting and visual connections within. The courtyard is also an exhibition space.

The second phase of DOX, Gulliver, was realised between 2014–2016 by architect Martin Rajniš. It is a horizontal tower. Built into the space of sky at the centre of the courtyard, it is an ambiguous airship, perhaps landing, crashing, rising or floating. A hidden steel construction wrapped in wood and plastic screens stands on two steel piers. Inside is a small stage and an open space dedicated for literature, reading and performances. It is a symbol of the surreal spirit hidden in the city blocks.

The third phase, DOX+, was realised between 2016 and 2018 by Petr Hájek. It is an extension into the neighbouring 6-storey building and courtyard. It includes a new auditorium, a rehearsal room, a café, administrative spaces and an extended courtyard. On the lower levels, the new auditorium and rehearsal room were built as semi-detached objects. The large auditorium is integrated into the internal circulation of DOX I, and provides over 200 seats. This new interior is in exposed concrete while its exterior is wrapped in cushioned, grey polymer sheets, giving it a soft and airy appearance. The roof of the auditorium is a sloped performance space covered in artificial grass. On the upper floors are a cafe and offices.

As a call for discovery and renewal upon an ambiguous threshold, Gulliver perhaps best symbolises the spirit of the contemporary project in Prague. It is a figure within and beyond, open yet bound. It provokes the possibility of something new, the possibility of something known, and the possibility of something other. In form and as symbol, it presents the openness and edginess of a place pivoting in space and time.

布拉格艺术馆
Kunsthalle Praha

地址：Klárov 5, 11800 Praha 1, Malá Strana
交通：Tram 27, 17, 2, Metro A, **Stop** Malostranská（3 分钟步行）
建造时间：2019—2022 年
建筑师：施德勒 - 塞科建筑事务所 (Schindler Seko Architects)

■ 新古典主义/ NEO-CLASSICAL
☐ 当代/ CONTEMPORARY

布拉格艺术馆是布拉格历史中心一个新的非营利文化空间，位于小城与莱特拉公园的交会处，马路对面就是通往布拉格城堡的大阶梯。这是一个由变电站改造的艺术馆，主要展出20世纪和21世纪的捷克和世界当代艺术作品，并为大众提供广泛的展览和教育服务。

原变电站由布拉格电力公司于1929—1932年建造，在1932年开始运营，主要服务于该地区电车和无轨电车线路运行的变电压、电流要求。该变电站建在文艺复兴时期的宫殿和18世纪军营的基础上，由建筑师维勒姆·克瓦斯尼卡以保守的新古典主义风格设计，来与小城的古典建筑群相协调。但同时，建筑内部的设施技术在当时处于非常先进的水平。该变电站以著名物理学家和气象学家、捷克技术大学教授瓦茨拉夫·卡雷尔·贝德里希·曾格尔的名字命名。随着变电技术的不断进步和设备升级，所需的设备空间日益小型化。从2000—2010年，整个变电站进行了现代化改造，变压器系统也搬到了地下室，只占据一小部分空间，因此建筑大部分空间仍然空置。2015年，曾格尔变电站被普迪尔家族基金会收购，改建成为今天的布拉格艺术馆。

改建设计由捷克本土的施德勒-塞科建筑事务所完成，旨在将工业建筑遗产的技术及空间复杂特征与当今社会、艺术和文化需求相结合，在尽可能保留建筑历史特征的基础上，融合不同时期的历史断面，通过新旧元素的细致整合，活化成为一个环境舒适并符合当代艺术展览最高标准的画廊空间，创造出一个激发灵感和想象、鼓舞人心的场所。新的功能需求包括了三个独立的展览空间、艺术品储藏和修复、会议、商店及餐饮等。

原有建筑外部的古典形态与历史环境相适应，但与内部空间的逻辑并不一致。建筑现代化的平屋顶、钢筋混凝土内部结构和机房被隐藏在新古典主义立面后面，从外部看不到内部高度现代化的技术特征。因室内地坪高出室外地面两米，因此从"正面"是无法进入的，进入变电站的入口是其侧面的一扇小门。由于结构材料老化及环境污染的问题，内部空间被完全拆除和替换，但外立面仍然是最重要的历史特征要素，其色调、窗框、表面纹理和构图都得到了细致的修复。建筑的特征要素和材料经过判定，被仔细地拆除、修复、编目和重新整合。一些独特的材料，如白色陶瓷绝缘体、铜线和20世纪70年代翻新时的混凝土面板，被重新构思为"化石"，再次加

工成新的建筑构件。拆除过程中产生的碎片和残留物被转化为新的循环材料。利用原建筑的石材和碎片合成特殊混凝土制品的同时,保留了其原有的颜色、纹理和品质。

建筑外部唯一的干预是入口坡道的重新设计。青铜色的新坡道面朝城堡和主干道,如抽象雕塑一般从地面伸出,街道的石头铺砌和景观纹理也一并延伸到坡道上来,室内外环境得到交融。游客由此进入一个"新"正面,该空间从上到下开放,形成一个连续的多层空间体量,实现了空间属性的真正转化。屋顶的露台面向城堡,又将建筑纳入了城市的历史环境。

从服务于城市交通的变电站到服务于城市文化的艺术馆,建筑完成了从提供城市运行动力到提供文化动力的转变。这个过程反映出这座城市的发展理念:建筑遗产的活化利用不仅要满足当代社会发展的需求,同时还要保存其"DNA",增强其场所精神。

… a former transformer station transforms dynamically, yet almost does not change.

While the Modern building electrified the city, its Contemporary renovation aimed to power the city in a different way. The objective was to insert an independent space for culture, as a house for the public collection of art and dialogue, in the heart of the historical city. In both a physical and symbolic sense it fuses the historical layers of the past through many scales. With its careful integration of new and old elements, it makes of the revitalised historical form a more authentic original.

Built as the Zenger Transformer Substation building from 1929 to 1932 by architect Vilém Kvasnička for the Electrical Company of the City of Prague, it was a state-of-the-art modern transformer station for the electrification of Prague's tramlines. While electrification transformed the inner workings of society, these Modern-era advances were too radical to be publicly displayed to a still adapting society. The building was to be masked. The Contemporary renovation, realised from 2017 to 2022 for the Pudil Family Foundation by Schindler Seko Architects, turned the facade inwards to become the internal generator itself.

The original building's flat roof, reinforced concrete interior and machine rooms were hidden from the outside behind a Neo-Classical facade, and one could not see the highly modern technical facilities within. While the morphology and articulation of the exterior responded to its historical surroundings and terrain, it did not correspond with its interior. The "front facade" was inaccessible, as the floor was two meters above ground level, and the original entry into the industrial building was a small technical door out of sight. By the 1970s, as the building technology advanced, the machinery was relocated to a miniaturised

facility in a single room in the building's basement and the building remained mostly empty.

The name Kunsthalle Praha itself reflects the memory of a multi-lingual, multi-cultural identity of the city and represents a layered space. This renewed, new-old building can be seen as a vertical section through the historical composition of the city. Built upon the foundations of a Renaissance palace and an 18th century military barracks, it rests upon the rocks of the earth through which the Brusnice stream still meanders into the Vltava river. The Modernist conception of form (appearing only at the free-standing columns on the northern facade) was otherwise encased within a Neo-Classicist image. This difference between the historical facade and the industrial interior was thematised in the Contemporary renovation.

Rather than as an applied image used to satisfy a conservative public, the facade became an integral part of the interior and its relation to the city. The compositional logic of the previously dislocated facade was used to produce a new spatial configuration as a raumplan. The new building includes rotating exhibitions of modern and contemporary art as well as educational activities, three separate exhibition spaces, one installed above roof level, an art gallery, technical rooms for storage and restoration of art, as well as meeting spaces for the public, a bistro and cafe. While the old interior was completely removed and replaced, the facade remained a free-standing artefact, with its colour tones, window frames, surface textures and composition restored in meticulous detail.

The renovation required other complex techniques of adaptive reuse as the building material was transformed at many scales. Not only was a provisional steel structure built from both sides to support the exterior wall as an additional basement level was added (30cm below the level of the Vltava river) and the new concrete structure installed. The 7 story building aimed to ensure the architectural specificity, its industrial and historical character, was not erased but rather transformed anew at the granular level.

An approach of "rescue archaeology" was used. Valuable historical elements and materials were carefully dismantled and restored, cataloged and reintegrated. Unique material substances, such as white ceramic insulators, copper wires, and concrete panels from 1970s renovations, were re-conceived as "fossils", re-processed into new tiles or tables. Debris and residues produced from the demolition were transformed into new up-cycled materials. Special concrete fabrication used stones and fragments from the original building to produce new material while retaining its authentic site-specific colours, textures and qualities. This digestive process at the scale of the building and the detail reflects a philosophy of the city: a conversion of historical substance into contemporary needs while retaining and enhancing the "DNA" and spirit of the place.

The sole new intervention on the exterior is the entry ramp on the primary, western facade looking towards the Castle. The bronze coloured figure is pulled out from the ground, bringing the stone paving of the street and the texture of the landscape to the building's elevated threshold. Visitors can now enter the "new" front facade above ground level into the centre of a space which opens both above and below into a contiguous multi-layered spatial volume. The urban terrace facing the Castle at roof level, finally accessible from within, links back into the city. In an advance from the representative balustrade of the Rudolfinum, the free-standing figures are now the people themselves.

新的未来

也许……这是又一个新的门槛。

冷战结束后,捷克共和国在地缘政治上实现从"东方欧洲"向"西方欧洲"的转型,经济在短暂的衰退后得到迅速地复苏,2006 年,捷克共和国被世界银行承认为"发达国家",并成为当今工业化程度最高的欧盟经济体之一。在新的千年,作为捷克首都的布拉格以其丰厚的历史文化沉淀和当代先锋潮流再次融合,并焕发生机,布拉格也因此重新成为欧洲的文化艺术之都之一。

布拉格过去一千年的历史及其建成环境是累积的、渐进的。各个时期都在此留下了印记,并极少遭受破坏,这是历史留给布拉格最丰厚的遗产。布拉格的历史也已经成为这座城市无可替代的、日益增长的价值。在对历史城市进行严格保护的前提下,布拉格的建成遗产可为公共利益以及社会文化服务;对城市遗产的保护可作为一种城市文化的推力,对当代社会审慎地进行文化塑型。与此同时,布拉格的新建筑也需要服务于公共利益和城市遗产保护。但是,作为旅游产业高度发达的城市,这座城市的遗产及历史是否仍会成为其日常生活和未来的一部分?在城市不断发展的过程中,新的生命如何与过去的痕迹形成共生关系?场所精神和历史叠加痕迹的价值是什么?同时,在全球化、数字化的浪潮中是否面临着同质化而失去自身特质的风险?这些都是现在的布拉格所面对的问题。希望新的城市繁荣不会轻易抹去布拉格已形成的千年积淀。

作为欧洲的几何中心,世界最美丽的城市之一,布拉格连续多样的建成环境和所形成的集体记忆理当继续在当今和未来发挥作用。过去的多种意识形态在今天继续发挥作用,但无法预知会产生哪些新的城市结构,以及它们如何与过去的城市结构相关联。或许,在不久的将来,城市景观和地标建筑之间会有新的联系;也许,会有新的桥梁、轴线、节点及公共空间的产生;或许,城市"中间"地区将发展成为新的次中心从而振兴城市核心;或许将看到艺术和文化的复兴以及市政当局的努力,一如过去,朝向共同目标对抗其他强大的阻力。对布拉格城市及其周边区域的不断扩展,要有一个指导性的规划意见(就像查理四世对于新城的

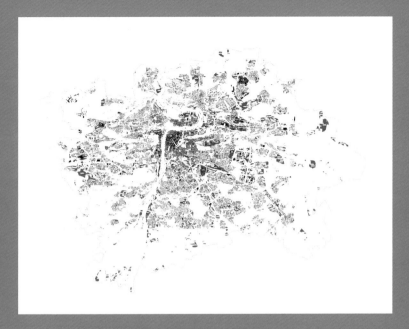

未来的布拉格
The Near Future City

规划一样），例如城市人口的不断增加而造成的住宅短缺、共产主义时期建造的住宅区改造和城市老工业区更新的应对策略等。在城市可持续发展的过程中，哪些建筑及环境更需要、更适合进行保护？可以兴建什么具有可持续品质的文化生产场所？这些都是今后的布拉格所面对的问题。

这表明，拼贴的布拉格仍然处于新复兴的开始，让我们相逢在 20 年后再看不断变化的布拉格吧。

THE NEAR FUTURE

... perhaps a different threshold ...

The history of the past thousand years in Prague and its architecture have been cumulative, progressive and disruptive, with the same forces continuing to play a part in this dynamic process of making and remaking the land and the people. The interplay — over time — among the elements we examined (land, people, markets, law and the sacred) may become less complicated and less productive if only one force (i.e. the market as law) were to overtake all the others (i.e. the sacred land of the people).

A great mixture of ideas and people, thriving markets and future potentials, indicate that Prague is, still and again, upon the cusp of a new revival. The Czech Republic is today the most industrialised EU economy by percentage, with the third most dense railway system in Europe. Advancing technologies bring new forms of possibility.

From the historical centres to outside the 14th century boundaries of the New Town, through and beyond the 19th century limits, into the interstitial zones between the outlying peripheries of the 21st century, public institutions, cultural infrastructure, multi-modal transport linkages and hubs of exchange, or simply houses, can make new landmarks of value within the structure of the near future landscape.

Prague should increase its population density by 4 times to achieve a sustainable socio-economic system. Therefore there is a large housing crisis, which requires not only homes but technical and cultural infrastructure. In many ways this circumstance may parallel the urban booms of the two golden eras of Prague, under Charles IV and Tomas G Masaryk, in the mid-14th century and early 20th century, when the civic authorities, the merchants and the people cooperated to produce a woven urban fabric with a sacred symbolic dimension.

The city is a collective memory in the making. Multiple ideologies of the past continue to play a role in the present, and we can not know which types of structures will be upcoming and how they will relate to what came before. One may wonder to what degree the legacy of the city and its history will remain a part of the everyday life of its future. Despite the variety in architectural styles of a future Prague, and the potential to profit from its growth, there must also be a guiding vision, overarching yet specific, (like that put forth by Charles IV) for the evolving city and its surroundings.

Recently, the city formed the Institute of Planning and Development. A new Metropolitan Plan was produced wherein the urban landscape is documented in extraordinary detail with its information accessible to the public. The objective of the near future is to make the city attractive to visitors and residents as a site of both heritage and new content. Collections of architects and investors are collaborating with the city as an investor as well, in public-private initiatives that also engage the input of the people. The upcoming developments and priorities intend to change the face of Prague significantly if they were to be completed.

There will be new infrastructural connections: another metro line, and railroads to the airport, completed ring roads for car transport, and new pedestrian bridges. New squares and parks will be built. Entirely new districts, over 50 times larger than Wenceslas Square, will emerge from the brownfields designated as "transformational areas" within the city centre alone. And new cultural spaces will be woven through the adaptive reuse of abandoned buildings of value.

While we looked at the architectural objects of the Contemporary time which bring hidden values out from within a rich fabric at an architectural scale, the near future operates at a much larger scale. This strategic planning integrates great complexity through multiple stakeholders, linking and stitching the landscape with connections into existing infrastructure. Unlike the earlier golden eras, when a smaller scale of land was owned by a larger group of people, today and in the near future, larger scales of land are owned by fewer groups of people. As such, the economic planning, scale of intervention and time scheduling of the near future development of the city will produce a specific dimension and character of its time. Only the distant future will show how time will have settled into place to have become new layers within the emerging landscape.

We will meet again in 20 years — at the threshold of a new Prague — when today's near future becomes history.

建筑名录

1. 高堡的圣马丁圆形小教堂	罗马风	—1070？
2. 犹太区老-新犹太教堂	早期哥特	1270？
3. 布拉格城堡内的圣维特大教堂	哥特（到当代）	1344—1929
4. 查理大桥	晚期哥特	1357—
5. 老城广场的旧市政厅	晚期哥特	1338—
6. 安娜王后的夏宫	早期文艺复兴	1538—1563
7. 施瓦岑贝格宫	晚期文艺复兴	1545—1567
8. 华伦斯坦宫和花园	早期巴洛克	1621—1630
9. 克拉姆-葛拉斯宫	巴洛克	1713—1719
10. 圣尼古拉斯教堂(小城)	早期和晚期巴洛克	1673—
11. 斯特拉霍夫修道院及其图书馆	巴洛克/洛可可	1140—1797
12. 城邦剧院	洛可可/新古典主义	1781—1783
13. 金斯基花园	新古典主义	1825—1831
14. 马萨里克火车站	复兴主义/新哥特	1841—1866
15. 国家大剧院	新文艺复兴	1868—1883
16. 鲁道夫音乐厅	新文艺复兴	1876—1884
17. 布拉格中央火车站	新艺术运动	1901—1909
18. 市民会馆	新艺术运动	1905—1911
19. 卢塞纳宫	新艺术运动和装饰艺术	1908—1921
20. 黑色圣母屋	立体主义	1911—1912
21. 亚德里亚宫	朗多立体主义	1923—1925
22. 展览会宫	结构主义	1925—1928
23. 拔佳鞋屋	功能主义	1927—1929
24. 马内斯美术馆	现代主义	1928—1930
25. 芭芭住宅区	现代主义	1929—1932
26. 缪勒别墅	现代主义	1930—1932
27. 耶稣圣心教堂	现代主义	1929—1932
28. 国际饭店	社会主义现实主义	1952—1956
29. 斯拉夫本笃会修道院（修复）	现代新哥特	1965—1968

30. 联邦议会大厦	新构成粗野主义	1966—1973
31. Máj 百货公司	新功能后现代主义	1972—1975
32. 布拉格会议中心	新功能粗野主义	1976—1981
33. 新舞台	粗野后现代主义	1977—1983
34. 日什科夫电视塔	新未来主义	1985—1991
35. 会跳舞的房子	当代	1992—1994
36. 金色天使大厦	当代	1999—2000
37. 皇家花园的柑橘暖房	当代	1999—2001
38. 鹿谷步道	当代	1999—2002
39. 欧洲大厦	当代	2002—2004
40. 国家技术图书馆	当代	2006—2009
41. DRN/ 民族大厦	当代	2012—2017
42. DOX 当代艺术中心	当代	2003—2018
43. 布拉格艺术馆	当代	2019—2022

INDIVIDUAL BUILDING TABLE

1. St. Martin's Rotunda at Vyšehrad Castle	ROMANESQUE	–1070?
2. Old-New Synagogue of the Jewish Town	EARLY GOTHIC	–1270?
3. St. Vitus Cathedral in Prague Castle	GOTHIC (THROUGH CONTEMPORARY)	1344–1929
4. Charles Bridge	LATE GOTHIC	1357–
5. Old Town Hall	LATE GOTHIC	1338–
6. Queen Anne's Summer Palace	EARLY RENAISSANCE	1538–1563
7. Schwarzenberg Palace	LATE RENAISSANCE	1545–1567
8. Wallenstein Palace and Gardens	EARLY BAROQUE	1621–1630
9. Clam-Gallas Palace	BAROQUE	1713–1719
10. St. Nicholas Church (Lesser Town)	EARLY AND LATE BAROQUE	1673–
11. Strahov Monastery and its Library	BAROQUE / ROCOCO	1140–1797
12. The Estates Theatre	ROCOCO/NEO-CLASSICAL	1781–1783
13. Kinský Folly and Garden	NEO-CLASSICAL	1825–1831
14. Masaryk Railway Station	REVIVALISM / NEO-GOTHIC	1841–1866
15. National Theatre	NEO-RENAISSANCE	1868–1883
16. Rudolfinum	NEO-RENAISSANCE	1876–1884
17. Prague Main Train Station	ART NOUVEAU	1901–1909
18. Municipal House	ART NOUVEAU	1905–1911
19. Lucerna Palace	ART NOUVEAU / ART DECO	1908–1921
20. House of the Black Madonna	CUBISM	1911–1912
21. Adria Palace	RONDO CUBISM	1923–1925
22. Trade Fair Palace	CONSTRUCTIVISM	1925–1928
23. Baťa House (Palace) of Shoes	FUNCTIONALISM	1927–1929
24. Mánes Union of Fine Arts	MODERNISM	1928–1930
25. Baba Residential Estate	MODERNISM	1929–1932
26. Villa Müller	MODERNISM	1930–1932
27. Church of the Most Sacred Heart of Our Lord	MODERNISM	1929–1932
28. Hotel International	SOCIALIST REALISM	1952–1956
29. Emauzy Convent	NEO-GOTHIC MODERNISM	1965–1968
30. Federal Assembly Building	NEO-CONSTRUCTIVIST BRUTALISM	1966–1973
31. Máj Department Store	NEO-FUNCTIONALIST POSTMODERNISM	1972–1975
32. Prague Congress Centre	NEO-FUNCTIONALIST BRUTALISM	1976–1981
33. New Scene of National Theatre	BRUTALIST POSTMODERNISM	1977–1983

34. Žižkov Television Tower	NEO-FUTURISM	1985–1991
35. Dancing House	CONTEMPORARY	1992–1994
36. Golden Angel	CONTEMPORARY	1999–2000
37. Orangery in the Royal Gardens	CONTEMPORARY	1999–2001
38. Deer Moat Passage	CONTEMPORARY	1999–2002
39. Palace Euro	CONTEMPORARY	2002–2004
40. National Technical Library	CONTEMPORARY	2006–2009
41. DRN / Národní Palace	CONTEMPORARY	2012–2017
42. DOX Centre for Contemporary Art	CONTEMPORARY	2003–2018
43. Kunsthalle Praha	CONTEMPORARY	2019–2022

参考文献
Reference

[1] 马奥尼.捷克和斯洛伐克史[M].陈静,译.上海:东方出版中心,2013.

[2] 伊贝林斯.19世纪末—21世纪初的欧洲建筑[M].徐哲文,申祖烈,译.北京:中国建筑工业出版社,2016.

[3] 陈翚,徐昊皓.20世纪先锋建筑的序曲:1910—1928年捷克立体主义建筑实践[M].北京:中国建筑工业出版社,2016.

[4] Chris van Uffelen, Markus Golser. Prague: The Architecture Guide[M]. Berlin: Braun Publishing, 2013.

[5] Jan Boněk. Cubist Prague[M]. Prague: Eminent Publishing House, 2014.

[6] DK Publishing. Prague 2017(DK Eyewitness Travel Guide)[M]. London: Dorling Kindersley Limited, 2016.

[7] Michael Kohout, Vladimir Šlapeta, Stephan Templ. Prague 20th Century Architecture[M]. Prague: Zlatýřez, s.r.o, 2008.

[8] Irena Fialová, Jana Tichá. Prg 2021 Contemporary Architecture[M]. Prague: Zlatýřez, s.r.o, 2007.

历史照片出处
Historical Picture Sources

建筑3、4、6、10、12、14、15、16、17、18、29
Pavel Scheufler. World of Prague[M]. Prague: Pražský Svět Publishing House, 2000.

建筑2、5、28、29、30、33、35、36
J.M.Lau. Prague Then and Now[M]. San Diego: Thunder Bay Press, 2007.

后记一

我第一次访问布拉格是在 2014 年深秋，没想到就此与这座美丽的城市结下了不解之缘。本书是我在多次访学布拉格，实地调研基础上的成果，也是对传统西方建筑史书中较少涉猎的中东欧建筑研究的扩展。可以说，布拉格建筑的丰富性、完整性和多样性在世界遗产城市中首屈一指，是建筑学人和建筑艺术爱好者欧洲旅行必不可缺的目的地之一。

本书的合作者埃兰·诺依曼·菲仕乐（Elan Neuman Fessler）是一名美捷双重国籍的犹太建筑师、建筑学教师。他本身的家族经历就是近现代捷克史的缩影。他的专业性，特别是对建筑哲理性的分析及图式语言让本书有了理论的高度。

感谢捷克的学者和朋友 Regina Loukotová、Klára Doleželová、Hana Benešovská、Babora Šimonová、Makata Mracková、Martin Stoss、Miroslav Vochta、Redek Toman、Jana Duero 等，布拉格的历史建筑众多，关于代表建筑的选定也很难取舍，最终入选的建筑也是与捷克同仁们多次讨论的结果。这本书也是中捷文化交流合作与友谊的结晶。

感谢中国驻捷克大使馆历任教育组负责人张振康、唐云、赵长涛老师，布拉格中华学校校长戴波老师对本书的支持。

感谢恩师常青院士对于我在中东欧访学及研究的支持，他对建筑遗产保护的诸多学术思想影响着本书的写作。同时也感谢江波、黄颖、祁学银、华耘、左琰、李海清、朱晓明、余平、沈奇、罗宁、江岱、刘杰、李辉、陈青长、李孟顺、王庆毅、朱建军、杨晓绮、张峥、刘晓卉、黄琪、曾庆璨、赵苏等师友对于本书的鼓励和支持。我们中有的曾经一起在布拉格穿行，或在中国探讨中欧建筑，你们的支持都是本书形成的动力。

还要感谢《建筑遗产》和 Built Heritage 编辑部的同仁们，文中很多内容与你们多次讨论。感谢本书的责任编辑及美编的努力让本书得以更好地呈现。

最后感谢我的家人，你们一直以来的理解和支持是我能安心研究写作最大的保障。

尽管本书从策划到完稿历经五年，但由于作者的认知和水平有限，书中还有不足之处，请大家批评指正（jj70s1@hotmail.com）。

2021.10.01

Afterword 1

I first visited Prague in the late autumn of 2014. I didn't expect to form an intimate bond with this beautiful city. Many visits to Prague and extensive field work culminated in this book. It extends previous research work on Central and Eastern Europe architecture, a topic less well covered in the literature on architectural history from the West. It can be said that the richness, integrity and diversity of architecture in Prague are second to none among world heritage cities. It is an essential city of interest for architects and architecture lovers travelling in Europe.

Elan Neuman Fessler, the co-author of this book, is a Jewish architect and architecture teacher holding dual US and Czech citizenship. The experiences of his own family are the epitome of modern Czech history. His professionalism, especially the analysis of architectural philosophy and schematic language, offers theoretical depth to the book.

I'm immensely grateful to my Czech colleagues: Regina Loukotová, Klára Doleželová, Hana Benešovská, Barbora Šimonová, Markéta Mráčková, Martin Stoss, Miroslav Vochta, Radek Toman, Jana Duero and other scholars and friends. Prague has many historical buildings, and it is difficult to choose particular representatives. The selection of buildings is the result of many discussions. This book is also a symbol of Sino-Czech cultural exchanges, cooperation and friendship.

I'm indebted to the successive heads of the education office of the Chinese Embassy in Czech Republic: Zhang Zhenkang, Tang Yun and Zhao Changtao; Dai Bo, President of Prague Chinese School, who offered their support for this book.

I'm also indebted to my tutor, Chang Qing, the Academician of CAS, for his support for my research in Central and Eastern Europe. His profound academic thoughts on architectural heritage conservation have influenced the writing of this book. At the same time, I also thank Jiang Bo, Huang Ying, Qi Xueyin, Hua Yun, Zuo Yan, Li Haiqing, Zhu Xiaoming, Yu Ping, Shen Qi, Luo Ning, Jiang Dai, Liu Jie, Li Hui, Chen Qingchang, Li Mengshun, Wang Qingyi, Zhu Jianjun, Yang Xiaoqi, Zhang Zheng, Liu Xiaohui, Huang Qi, Linck Zeng, Coco Zhao and other professors and friends for their encouragement and support for this book. Some of us have traveled together in Prague or discussed Central European architecture in China. Your support is the driving force for the formation of this book.

I would also like to thank my colleagues in the editorial department of *Heritage Architecture* and *Built Heritage* for many discussions with you. Thanks to the efforts of the editor and graphic designer, this book has found its appropriate presentation.

Finally, this book would not have been possible without my family. Your understanding and support are the greatest guarantee for me to study and write at ease.

Even though the writing of this book has spanned five years from planning to completion, room for improvement and polishing remains, not least because the author is still a learner seeking new insights. Please feel free to send errata, comments or criticism to jj70s1@hotmail.com.

Pu Yijun
2021.10.01

后记二

透过我母亲小时候的照片，我看到了布拉格的过去，在那个年代我的外祖父母穿着套装，戴着礼帽。近 80 年了，一个世纪，甚至一千年过去了，照片背景中的城市与建筑仍然像今天我们看到的一样伫立在那里。也许时间是线性的，但当我们在其中徘徊时，时间会不断与它的过去交叠。如果我们换个角度看世界，就能看到之前从未看过的景象：所有事物都被放置在恰当的位置，并相互重叠。每一幢建筑都是嵌入空间的人造物，每一个场所都是隐藏过去的棱镜。它们以时光流逝为抽象的边界，勾勒出一幅群像。这座城市在一代代人的时间里向自身投射过去的剪影，经年累月地堆积着它的神秘。这些无尽的、静止的轨迹构成了今天隐匿、平坦而无穷的景观，其深度构筑出唯一的空间。这就是共通时间的奇妙之处：它看似消失，却又无处不在。

本书是一个个连续展开的图像，形成文本，并以叙述的方式带读者领略一个地方所展现的世界。在时光旅行中我们无需移动，这座城市和它隐藏的维度穿过我们，揭示其过去的样貌。我们用没有文字的史诗在时光中营造故事。这首由音符、声音和线条组成的交响曲锻造出我们已存在潜意识里的身体。布拉格作为城市，作为符号，作为"门槛"，它的最终信息永不为人知。我们会渴望与这些过去的声音和形象对话，或者有一种将它们带回现在，并作为我们的一部分而不是历史的一部分来保存的动力。这种动力无关怀旧，因为未来也不是我们的，它取决于我们记住了什么。

这不仅仅是一本是关于布拉格历史及建筑的书，它更是一条带领我们穿行于凝固的思想史中的路，而我们仍生活于其中。这段思想史中无数的想法以具体的形式呈现在现实世界时，就成为了如今我们仍旧居住其中的建成环境。在此，我要感谢同济大学出版社，感谢我的中国好友蒲仪军和我一起踏上了这趟穿越千年的旅途，还要感谢布拉格建筑学院的同事们，特别是 Michaela Janečková 在成书过程中给予我的宝贵意见。还要感谢我的家人，无论是过去还是未来，是他们让我找到了自己的位置。我希望这本书能帮助我们揭开当下所看到表象，为那些隐匿在表象之下的更深刻的东西在我们的头脑中留下一席之地，让今天再编织成为未来的一个组成部分。

埃兰·诺依曼·菲仕乐

2021.10.28

Afterword 2

    I can see in the photographs of my mother as a child, my grandparents of an other era dressed in suit and hat. Nearly 80 years, or a century, or a millennium past, they still stand in full presence as if it were today. Perhaps time is linear, but it crosses itself repeatedly as we wander in its midst. I can also see the photographs which were never taken, if I look at the world in a certain light. Each is positioned in its proper place, all overlapping. Every instance is an artifact embedded in space, and every place is a hidden prism of a past. They compose a multitude bounded by an imaginary line which is the passage of time. The city accumulates its mysteries as it casts shadows upon itself over generations. These unending and unmoving trajectories constitute the unseen, flat and infinite landscape within which the only space is depth. This is the magic of shared time: it has gone and it has gone nowhere.

    This book is a sequential unfolding of images, formed as text, and a narrative glimpse into the world of a single place. We do not need to move to travel through time. The city and its hidden dimensions pass through us revealing what is to be what was. We build stories in time, with each epic but a single word. This orchestral work of notes and voice and line forge that subconscious body in which we are all present. As city, as symbol, as threshold, its final message is never known. We may find desire to speak with these sounds and figures of the past, or a drive to retrace them into the present and, without nostalgia, to preserve them as a part of us rather than of history. As the future is also not ours, it depends upon what we have remembered.

    This book is not quite one of history or of architecture. It is a passage through the history of thought built as form within which we still live. I would like to thank Tongji University Press for commissioning this book, my dear friend Pu for embarking on this journey of a thousand years with me, and additionally my colleagues from ARCHIP, especially Michaela Janečková for her comments along the way. I would also like to thank my family, past and future, for positioning me where I have found myself. It is my hope that this book will erase the images of the present as we see them to allow for those beneath the surface to take their place in our minds, and for today to be conceived in turn as yet another layer in the composite to come.

2021.10.28

图书在版编目（CIP）数据

布拉格建筑地图 / 蒲仪军，（捷克）埃兰·诺依曼·菲仕乐 (Elan Neuman Fessler) 著 . -- 上海：同济大学出版社，2022.5
（城市行走书系 / 江岱，姜庆共主编）
ISBN 978-7-5765-0211-4

Ⅰ . ①布… Ⅱ . ①蒲… ②埃… Ⅲ . ①建筑物—介绍—布拉格 Ⅳ . ① TU-095.24

中国版本图书馆 CIP 数据核字 (2022) 第 090008 号

布拉格建筑地图

蒲仪军　　[捷] 埃兰·诺依曼·菲仕乐　　著

出 品 人　金英伟
责任编辑　金　言
责任校对　徐逢乔
书籍设计　张　微
出版发行　同济大学出版社 www.tongjipress.com.cn
　　　　　（地址：上海市四平路 1239 号　邮编：200092　电话：021–65985622）
经　　销　全国新华书店
印　　刷　上海雅昌艺术印刷有限公司
开　　本　710mm×1092mm　1/32
印　　张　9.25
字　　数　207 000
版　　次　2022 年 5 月第 1 版
印　　次　2022 年 5 月第 1 次印刷
书　　号　ISBN 978-7-5765-0211-4
定　　价　88.00 元

本书若有印装问题，请向本社发行部调换
版权所有　侵权必究